ADVANCED ANALYSIS
WITH THE SHARP 5100
SCIENTIFIC CALCULATOR

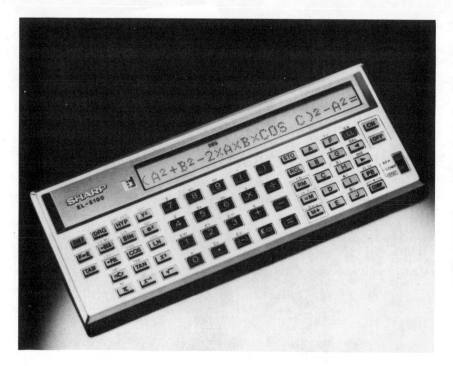

The Sharp 5100 keyboard is the SHAPE OF THE FUTURE. Small ASCII keyboards and alphanumeric dot matrix displays have been developed for the new hand-held computers now on the drawing boards of most computer companies. Many of these designs are in the horizontal keyboard format. The combination of keyboard shape and display is ideal for comfort, speed and convenience. Sharp decided to bring these benefits to you now. The 5100 scientific calculator has both the dot matrix display and the miniaturized ASCII keyboard shape.

ADVANCED ANALYSIS WITH THE SHARP 5100 SCIENTIFIC CALCULATOR

J. M. SMITH
JMSA Systems Research and Analysis

A Wiley-Interscience Publication
JOHN WILEY & SONS, New York · Chichester · Brisbane · Toronto

Library of Congress Cataloging in Publication Data

Smith, Jon M   1938–
    Advanced analysis with the Sharp 5100 scientific
calculator.

    "A Wiley-Interscience publication."
    An adaptation of the author's Scientific
analysis on the pocket calculator.
    Includes index.
    1. Sharp 5100 (Calculating-machine)
2. Numerical analysis.   I. Smith, Jon M.,
1938–   Scientific analysis on the pocket
calculator.   II. Title.

QA75.S554   510'.28   79-22505
ISBN 0-471-07753-4

PREFACE

This book is written for all who use a Sharp 5100 scientific calculator. The Sharp 5100 is a new and powerful computing instrument. When utilizing it in conjunction with the most modern numerical methods, the user has at his or her fingertips formidable computing power for scientific research, engineering design, and advanced mathematical analysis.

This book is an adaptation of the bestseller *Scientific Analysis on the Pocket Calculator*, 2nd edition (1977), published by John Wiley & Sons. It emphasizes numerical methods that are

- particularly suited to the Sharp 5100 scientific calculator;
- important for advanced scientific and engineering analysis.

Although the calculations are illustrated for use on the 5100 machine, the mathematical material is sufficiently general that it may also be used for calculations on any of the Sharp scientific machines, including those presently on Sharp's drawing boards.

Examples have been developed to emphasize the new computing power available to an analyst with a 5100 as a result of the unique algebraic entry capability designed into this new calculator. The breakthrough in function evaluation through alphanumeric entry of formulas is that the analyst is freed from the need to learn the calculator's machine language or special key stroke sequences to solve problems. This is a significant improvement over calculators using the algebraic operating system (AOS) and reverse polish notation (RPN) calculator languages. There are some who would argue that it is fun to program your calculator in machine language. The Sharp 5100 makes this an unnecessary step and allows direct entry of an algebraic equation and then, with its compiler, determines what would be the equivalent of a sequence of key strokes.

The Sharp 5100 is a major advance in pocket calculator computing capability and is one of the "new generation" of calculators that will lead the way to tomorrow's *pocket computers*. As an owner of a Sharp 5100, you are to be congratulated for staying current in this fast-moving field of personal computing equipment.

J. M. SMITH

Washington, D.C.
October 1979

CONTENTS

ADVANCED ANALYSIS WITH THE SHARP 5100 SCIENTIFIC CALCULATOR

CHAPTER 1

THE SHARP 5100 SCIENTIFIC CALCULATOR

1-1 INTRODUCTION

The Sharp 5100 scientific calculator is ideally suited to advanced mathematical analysis involving complex and sophisticated calculations. It is not, however, necessary to learn the 5100's "machine language" to program the Sharp 5100. All that is required is to program an algebraic function into the Algebraic Expression Reserve (AER) memory in alphanumeric form, just as you would find it in a handbook or textbook, and your 5100 will compile the expression into the appropriate sequence of operations in the machine's language to properly execute the algebraic function. If your check problem does not check out, all you have to do is proofread the formula and make corrections using the *insert* and/or *delete* keys.

But there is more: the Sharp 5100 is capable of much more than simply solving algebraic equations. While performing algebra on the 5100 in alphanumeric format is certainly a time-saver, the 5100 will also solve differential equations, simultaneous equations, and implicit equations, to name a few of its more advanced capabilities. This is possible because the 5100 can also be programmed to store and retrieve data from program addressable memory registers. Once a result is computed, it can be stored in memory and recalled for use in the next step of the calculation or in another iteration through the same algebraic formula. This permits the 5100 to solve problems using difference equations, recursion formulas, and many methods of successive approximations. Said slightly differently, with the proper numerical methods the Sharp 5100 becomes a tremendously powerful tool for conducting advanced numerical analysis. With its algebraic entry method, the 5100 is particularly suited for using the modern micronumerical methods presented in this book. Whether you are an engineer, mathematician, architect, physicist, chemist, or statistician, the

1

5100 will be an asset because you can program this new machine in the algebraic form you want and are most familiar with. It is not necessary to translate your equation into a form that is machine compatible and then either memorize the key strokes necessary for numerical evaluation or program the machine to execute the proper key strokes automatically.

There are certain elementary concepts in numerical analysis that are both entertaining to learn and important to understand before proceeding to advanced analysis. The concepts are easy to comprehend and are basic to calculating with numbers. They show how to set up a problem so it may be solved in the most accurate and efficient manner:

1. With the least chance for error.
2. With the least sensitivity to error propagation.

It may surprise you to learn that the *number system* used by any calculator can significantly affect your calculations. This is particularly true when you are using very large or very small numbers or when an intermediate calculation results in a very large or very small number. In what follows, we discuss the practical effects of the 5100's treatment of small and large numbers and its effect on computing accuracy. We also address methods of rearranging mathematical expressions to minimize error generation and propagation.

1-2 THE FLOATING-POINT NUMBER SYSTEM

The Sharp 5100 performs calculations in the floating-point number system. *Floating-point numbers* have a decimal point that moves so as to retain the most significant digits in any calculation. What is surprising to many is that the field of numbers in the 5100 is "bunched" around zero. To visualize this interesting phenomenon, let us examine the distance between numbers in the 5100's floating-point number system. First, note that the distance between zero and the smallest nonzero positive number is 0.0000000001. Next notice that the distance between the largest positive number 9999999999 and its next neighbors is 1. We see here that the distance between the numbers in the floating-point number system is not uniform. In fact, the distance between numbers varies from 1 to 0.0000000001 (factor of 10^{10}), depending on whether the number in the display is very small or very large. Said another way: in the field of floating-point numbers, the distribution (distance between numbers) is not uniform. In fact, there are as many numbers grouped between 0 and 1 as there are between 1 and the full register size 9999999999. But wait:

although the absolute difference varies significantly in floating-point numbers, the percentage difference remains fixed. As used in this book, *percentage difference* is the ratio of the difference between two consecutive numbers divided by the larger of the two. "So what?" we might ask. Well, for most engineering and scientific analysis, percentage difference and percentage error are usually the measure of accuracy of most interest, even when doing large-number calculations. Only infrequently do we pursue answers to the question, "What is the absolute error in this calculation?" In this sense, then, the floating-point number system is better suited to engineering and scientific analysis, and particularly to analysis with large numbers. Any errors introduced by the use of the 5100 number system will result in a fixed percentage error over the entire range of numbers.

In the Sharp 5100 the floating-point number system is extended by powers of 10, permitting the positive floating-point numbers to range from 10^{-99} to $9999999999 \times 10^{99}$. Interestingly, this even further bunches the floating-point numbers in the neighborhood of zero. Because of this grouping property, the absolute errors are smaller for calculations with numbers between 0 and 1 than for numbers between 1 and the full range of the calculator. So, if you are doing analysis with very large numbers where absolute error is important, be careful to determine through independent means the effect of each calculation on the accuracy of the result.

1-3 THE EFFECT OF OVERFLOW ON FUNCTION EVALUATION

Many good books on numerical analysis or computer operations present the equations for propagating relative or absolute error through a series of calculations. The formulas given for computing the error in the end result of a series of calculations, though accurate, are somewhat complex. It is our concern here to first learn to work within the limitations of the 5100's computing capability and to understand the calculator's impact on the generation of error that gets introduced into a problem. Then we study methods and techniques for getting around these problems.

Most calculations affect the range of the variables in a calculation through their treatment of both overflow and underflow. The Sharp 5100 was carefully designed to eliminate the underflow effect and has only an overflow effect. When a number exceeds the largest number in the 5100, the display is set to indicate an overflow error ....0..... When a calculation results in a number smaller than the smallest number in the 5100's field of numbers, the number is set to zero. Replacing an underflow by zero seems intuitively more reasonable than setting the calculator to show an underflow error as is done with some calculators. Now let us get familiar with the effects that underflow can have on a simple calculation. It is interesting to find that calculating the inverse of e^{-228} will not give the same result as calculating e^{228} directly. The reason is that e^{-228} is set equal to zero and thus the inverse is undefined, while e^{228} is within the 5100's field of numbers and thus calculable.

$$\frac{1}{e^{-228}} = \frac{1}{0_{underflow}} \rightarrow \text{undefined} = \ldots\ldots\ldots 0\ldots\ldots$$

$$e^{228} = 1.045061560 \times 10^{99}$$

This number system "end effect" can lead to some very practical limitations on the range of variables for which a function can be evaluated. Table 1-1 shows the effect of the 5100's overflow on the range of the function $x^5 e^x/(e^x - 1)$.

The function is written in two ways in the table, one way favoring underflow and the other way favoring overflow. That is, the function in the first column will eventually overflow the calculator's field of numbers because of the evaluation of $x^5 e^x$, while the function in the second column will eventually underflow the calculator's field of numbers because of the evaluation of e^{-x}. The table shows that the range of the variable x for which the function can be evaluated is dramatically limited for functions that overflow before they underflow. In fact, a function written in a form that will underflow before it overflows can explore a range of the argument that is often orders of magnitude greater than the same function written in a form that overflows before it underflows. In general, the Sharp 5100 favors functions written in the form that will underflow first. Were the underflow error included in the design of the Sharp 5100, x would only range between zero and about 300. By replacing an underflow with zero, the 5100 allows x to range from 0 to 1×10^{20}.

Table 1-1   The Effect of Overflow and Underflow on the Range of Function Evaluation

x	$x^5 e^x/(e^x - 1)$	$x^5/(1 - e^{-x})$
1	1.581976707	1.581976707
10	1.000045407×10^5	1.000045407×10^5
100	1.0×10^{10}	1×10^{10}
200	$3.200000023 \times 10^{11}$	$3.200000023 \times 10^{11}$
202	$3.363232171 \times 10^{11}$	$3.363232170 \times 10^{11}$
203	$3.447308812 \times 10^{11}$	$3.447308812 \times 10^{11}$
204	Overflow	$3.533058571 \times 10^{11}$
9×10^{20}	Overflow	$5.904899997 \times 10^{99}$
1.0×10^{20}	Overflow	Overflow

1-4 ROUNDOFF ERROR AND THE ACCURACY TRAP

There is a trap that anyone using a calculator can fall into, so watch out! Remember this formula

$$p_{\text{lost}} \leqslant 1 + \log_{10} N$$

This formula says (in words) that the number of places past the decimal point *you will lose* (p_{lost}) after performing N arithmetic operations (additions, subtractions, multiplications, and divisions) will, at worst, be equal to 1 plus the logarithm (to the base 10) of N. In other words, if you know the parameters in a problem to three places past the decimal point and you perform 10 arithmetic operations on the parameters, then *at worst* you are entitled to quote your result only to one place past the decimal point.

This in itself does not seem too bad; but it can be, and regularly is. Here is how it happens. Suppose we start with the two numbers 1.001 and 1.000, whose accuracy we will say is known to three places past the decimal point. The most significant digit in each of these numbers is one place before the decimal point. Suppose further that we are required to calculate the difference between these two numbers as $1.001 - 1.000 = 0.001$. We see that the most significant digit in the difference is now in the third place past the decimal point. This is very important—*the most significant digit has jumped from one place before the decimal point to three places past it*. The question is: "What can we say about the accuracy of this calculation?" When we started, each number in the calculation was known to three places past the decimal point. Now, applying the accuracy formula, we find

$$p_{\text{lost}} \leqslant 1 + \log_{10}(1)$$

which gives

$$p_{\text{lost}} \leqslant 1$$

We have lost one of the places past the decimal point in this single calculation, so that *at worst* we know the difference to only two places past the decimal point. This means we do not really have any reason to be confident in what is now the most significant digit. We do not really know if the difference is 0.001 or 0.00001, or even smaller, approaching an infinitesimal.

In general, we encounter the trap when taking the difference between two numbers of comparable magnitude or when taking the ratio of two numbers of comparable size. The trap is to assume that because the 5100 will calculate numbers with a 10-digit mantissa the accuracy of a number will be maintained through numerous arithmetic operations (calculations). Of course this is not true in general and particularly when calculating ratios and differences where the result of the calculation will move the least significant digits forward toward the most significant digits. Finally, keep in mind that when the number of places lost moves forward of the number of places required for accurate calculation of the most significant digits, the calculation result can be meaningless.

1-5 REARRANGING EXPRESSIONS TO MINIMIZE ERROR

Now let us examine the functions to be evaluated and how they may be written in forms where loss of accuracy due to the subtraction of two almost equal-sized numbers is minimized.

There are a number of "tricks" for accurately calculating the difference between two numbers that are of comparable magnitude. One general technique exists that can resolve many problem situations. Consider the function

$$h(x) = f(x+\epsilon) - f(x)$$

The numerical evaluation of $h(x)$ can propagate error forward into the leading significant digits. This function can be modified as

$$h(x) = \{f(x+\epsilon) - f(x)\}\left\{\frac{f(x+\epsilon) + f(x)}{f(x+\epsilon) + f(x)}\right\}$$

$$h(x) = \frac{f^2(x+\epsilon) - f^2(x)}{f(x+\epsilon) + f(x)}$$

This is a general equation that can, for algebraic and certain transcendental functions, transform the difference of two neighboring numbers into the ratio of sums of the numbers capable of being evaluated accurately. For example, if

$$f(x) = x^2$$

then

$$h(x) = \frac{(x+\epsilon)^4 - x^4}{2x^2 + 2x\epsilon + \epsilon^2} \cong \frac{4x^3\epsilon}{2x^2(1 + \epsilon/x)} \cong 2x\epsilon$$

For another example, consider the function

$$f(x) = \sin(x+\epsilon)$$

Then

$$h(x) = \sin(x + \epsilon) - \sin(x) = 2\cos\left(x + \frac{\epsilon}{2}\right)\sin\left(\frac{\epsilon}{2}\right)$$

which for small ϵ (but not necessarily small x) is

$$h \cong 2\left\{\cos\left(x + \frac{\epsilon}{2}\right)\right\}\left\{\frac{\epsilon}{2}\right\} \cong \epsilon\cos\left(x + \frac{\epsilon}{2}\right)$$

With regard to other techniques, Hamming makes the interesting observation that what appear to be a large number of tricks to reformulate a function to handle its finite difference are really not new to the analyst. They are exactly the same methods used in calculus to derive the function's derivative. We can see this from the definition of the derivative

$$\lim_{\Delta x \to 0}\left\{\frac{\Delta y}{\Delta x}\right\} = \lim_{\Delta x \to 0}\left\{\frac{f(x + \Delta x) - f(x)}{\Delta x}\right\}$$

As a final resort to avoiding subtraction of nearly equal-sized numbers, most functions can be series expanded or approximated with different types of series for the interval of interest. Then $h(x)$ can be formed and modified as before to get around the subtraction problem.

An approach that works surprisingly well for certain functions (see Example 1-4) is to use the mean value theorem of differential calculus, where

$$f(b) - f(a) = (b - a)f'(\theta), \qquad (a < \theta < b)$$

As an example of the application of the mean value theorem, let us compute

$$h(x) = \sin(x + \epsilon) - \sin(x)$$

where $x + \epsilon$ is not necessarily small. Using the mean value theorem, we find

$$h(x) = [(x + \epsilon) - (x)]\cos(\theta) = \epsilon\cos\theta$$

for

$$x + \epsilon > \theta > x$$

**Table 1-2 Commonly Used Difference Equations
in Functional Evaluation**

$$\Delta e^x = e^x(e^{\Delta x} - 1)$$

$$\Delta \ln(x) = \ln\left(1 + \frac{\Delta x}{x}\right)$$

$$\Delta \sin(2\pi x) = 2\sin(\pi \Delta x)\cos\left[2\pi\left(x + \frac{\Delta x}{2}\right)\right]$$

$$\Delta \cos(2\pi x) = -2\sin(\pi \Delta x)\sin\left[2\pi\left(x + \frac{\Delta x}{2}\right)\right]$$

$$\Delta \tan(2\pi x) = \sin(2\pi \Delta x)\sec(2\pi x)\sec(2\pi x + 2\pi \Delta x)$$

The difficulty is in selecting the value of θ that will accurately compute $f'(x)$; that is, in selecting θ that produces less error than would be produced by the propagation of error into the most significant digits. The author knows of no method for effectively estimating θ to ensure accuracy greater than is given by taking the difference itself. However, the midvalue interval is an obvious possibility. In this case, we find

$$h(x) \cong \epsilon \cos\left(x + \frac{\epsilon}{2}\right)$$

This method is of questionable value for precision evaluation except when θ can be determined either by experiment or by analysis. The equation is useful, however, for computing the extreme values of the difference by using the expressions

$$\epsilon \cos(x + \epsilon)$$

$$\epsilon \cos(x)$$

on the interval of the calculation.

A few commonly used difference equations for circumventing large errors in taking the difference between nearly equal values of popular transcendental functions are tabulated in Table 1-2. These difference equations can be programmed directly as written in the algebraic expression reserve mode of the 5100.

1-6 SIMULTANEOUS EQUATIONS

If you have your 5100 handy, set it up with the following expression in the AER*:

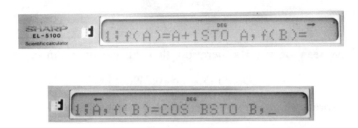

Then set the calculator to read angles in radians and let A = 0 and B = 0.5. Now repeatedly press the *COMP* key. (Remember, when "A?" appears in the display and no new "A" is input, the calculator will use the last contents in the "A" memory register when the *COMP* key is pressed again.) You will find the following sequence of numbers showing up in the display register:

	Ans 1	Ans 2
	Number of Iterations (A)	$\cos B$
	1	0.877582562
	5	0.768195831
	10	0.735006309
	20	0.739006780
	49	0.739085134
	50	0.738085133
	51	0.739085133

To what is this sequence of numbers converging? Said differently, what

---

*This equation can be rewritten in a form more convenient for the analyst:

$$1; \quad A + 1 \text{ STO A}, \quad \cos B \text{ STO B}$$

In this form pressing the *COMP* key once will calculate $A_n = A_{n-1} + 1$ and store A_n in the A memory register; pressing the *COMP* key a second time will calculate $B_n = \cos B_{n-1}$ and store B_n in the B memory register. B_n will be seen in the display register. To read A_n, press RCL A.

problem is being solved when a function is repeatedly evaluated using the results for the argument in the next iteration? The answer has powerful consequences as well as being interesting and of practical value. The number 0.739085133 is the solution to the simultaneous equations:

$$y = x$$

$$y = \cos(x)$$

We have seen, on the 51st iteration, that this equation converges as:

$$0.739085133 = \cos(0.739085133)$$

In other words, we have found the value of x where, numerically

$$x = \cos(x)$$

Now, when we started the calculation, the values of x were such that

$$x \neq \cos(x)$$

In our first key stroke, we calculated:

$$y_1 \approx \cos(x_0)$$
$$x_1 = y_1$$

Based on this, we could make the second approximation as:

$$y_2 \approx \cos(x_1)$$
$$x_2 = y_2$$

By repeating these iterations n times, we are implementing the finite version of

$$\lim_{n \to \infty} (x_n) = \cos(x_{n-1})$$

To summarize, when a function on your 5100 is iterated repeatedly using the AER mode and the sequence of display numbers converges, the result is the solution to the simultaneous equations:

$$y = x$$

$$y = f(x)$$

This is important from a practical point of view, because many engineering problems are iteration problems using implicit functions, although usually not quite so simple as our example. But we can extend the technique. Suppose you want to solve the tougher set of simultaneous equations:

$$y = h(x)$$
$$y = f(x)$$

For example,

$$y = x^2$$
$$y = \cos(x)$$

Then program your 5100 as follows:

Then store 0 (zero) in A and repeatedly press the *COMP* key until the process converges. In this case, you will find $x = 0.824132312$ in 26 iterations. This is one of the values of x where $x^2 = \cos(x)$.

Another example would be for

$$y = \cos(x)$$
$$y = \tan(x)$$

In this case, you would rewrite x to be:

$$x = \tan^{-1}(\cos(x))$$

and program it in the 5100 AER mode as*

and after 21 iterations you will find $x = 0.666239433$.

---

*Or 1; $\tan^{-1}\cos A$ STO A.

Will it work all the time? No. For example, if we attempt this technique with

$$x = \cos^{-1}(x)$$

we find the process will not converge. Similarly, if we try to solve the equation

$$\sin(x) = \cos(x)$$

by solving the equation

$$x = \sin^{-1}(\cos(x))$$

we find the solution to be a neutrally stable oscillation.

Finally, the solution of the equation:

$$e^x = x$$

simply goes unstable. Do you see why? This is something of a trick question. Have you found the answer? The reason is that $y = e^x$ and $y = x$ do not cross and form an intersection where x satisfies both functions simultaneously.

In general, there are convergence difficulties to contend with for many keyboard functions. Fortunately, however, they are easy to observe for most practical problems. Either the solution stops because of an overflow or an undefined argument, or the process does not converge.

The examples used here to illustrate the solution of simultaneous equations are purposely simple and use only a few keyboard functions. This numerical method can handle unusually complicated equations involving a great many of the keys (functions) on your 5100.

Interestingly, this problem of solving simultaneous equations is but one case of the more general problem of finding the zeros of a function. Note that simultaneity of

$$y(x) = z(x)$$

can be written as

$$y(x) = z(x) = 0 = Q(x)$$

and there are many useful numerical methods for solving this general problem.

1-7 RELATIVE ERROR IN APPROXIMATION FORMULAS

Approximation functions written in a form that minimizes relative error rather than absolute error have fewer terms and thus fewer arithmetic operations than approximating functions that minimize absolute error. Although this is well known to the experienced analyst, and seems quite rational to the practical analyst, we still find prevalent use of "absolute error" as an accuracy criterion in numerical analysis with approximation formulas. For example, calculating e^{-x} over the range 0 to 3, to 1 part in 10^3 using a Taylor series expansion about 0, requires on the order of 12 terms in the series approximation polynomial. That is, the contribution made by the 13th term in the Taylor series expansion of e^{-x}, when $x = 3$, is something less than 10^{-3}. However if, more reasonably, we require that the relative error be 1 part in 10^3, only 9 terms are needed in the series. The absolute error criterion requires 30% more terms than what is usually required for engineering analysis. In general, when deriving approximation formulas, it is important to decide what kind of error is important to the problem being solved and use approximation methods that provide the appropriate accuracy. Too often the approximation is laboriously long and overly accurate for the purpose.

1-8 NESTED PARENTHETICAL FORMS

Many functions of interest to engineers can be written in a power series. These series can be generated using Taylor's theorem, Maclaurin's theorem, Chebyshev polynomials, and so on. Often they arise when empirical data are "curve fit" with a power series (polynomials). When written in "standard" form, they take the following appearance:

$$f(x) = a_0 + a_1 x + a_2 x^2 + a_3 x^3 + \cdots + a_n x^n + \cdots$$

If we were to evaluate this polynomial as written, we would compute each term in the series and then sum them up to determine $f(x)$. The number of arithmetic operations would equal

$$6n^2 + 16n$$

for a 10-digit data entry as found with the 5100 machine. Notice that the number of operations goes as the square of the order of the polynomial. By rewriting this equation in the form

$$a_0 + x(a_1 + x(a_2 + x(a_3 + \cdots + x(a_{n-2} + x(a_{n-1} + a_n x)) \cdots)))$$

the number of operations is reduced to

$$11(2n-1)$$

We see then that when the polynomial is written in nested parenthetical form, the number of operations is only proportional to the order of the polynomial. Since each operation takes a finite amount of time, the fastest execution of polynomials on the Sharp 5100 scientific calculator is a-chieved when the polynomial is written in nested parenthetical form. This becomes important when doing iterative calculations manually by re-peatedly pressing the *COMP* key.

Example 1-1 Let us use the 5100 to investigate the difference between absolute error and relative error, as well as the best way to program polynomials (nested parenthetical forms) on the 5100. Evaluate $\ln(0.9)$ using the fifth-order truncated Taylor series expansion of $\ln(1+x)$ in the neighborhood of $x = 1$.

$$\ln(1+x) \cong x\left(1 - \frac{x}{2}\left(1 - \frac{2x}{3}\left(1 - \frac{3x}{4}\left(1 - \frac{4x}{5}\right)\right)\right)\right), \quad |x| < 1$$

Now

$$1 + x = 0.9$$
$$\therefore x = -0.1$$

Then, in the AER program

1; $f(A) = ((((-4 \times A \div 5 + 1) \times -3 \times A \div 4 + 1) \times -2 \times A \div 3 + 1) \times -A \div 2 + 1) \times A$, STOB $\blacktriangleright$
2; $f(A) = LN(1 + A)$, STOC $\blacktriangleright$
3; $f(B,C) = B - C$ $\blacktriangleright$

Accuracy considerations over a broader range of x are given in Table 1-3.

For those who are interested in the difference in speed for calculating this table using standard form polynomials, rewrite this series in standard form and program it to give Answer 1 with the nested parenthetical form programmed to give Answer 2. You will find the difference in time between the two forms to be so great that you will be able to time it with your stopwatch. Use nested parenthetical forms where possible when doing manual iteration with the 5100.

Table 1-3   Accuracy of the Fifth-Order Taylor Series Expansion of ln(1 + x)

(A)		2; Ans 2	1; Ans 1	3; Ans 1	$100 \times (B-C) \div C$
				Absolute	Relative
$(1+x)$	x	$\ln(1+x)$	$x[1-x/2(1-\cdots)]$	Error	Error (%)
0.9	-0.1	-0.10536052	-0.10536033	-0.00000018	00.000173
0.8	-0.2	-0.22314355	-0.22313067	-0.00001288	00.005774
0.7	-0.3	-0.35667494	-0.35651100	-0.00016394	00.04596194
0.6	-0.4	-0.51082562	-0.50978133	-0.00104429	00.20443190
0.5	-0.5	-0.69314718	-0.68854167	-0.00460551	00.80241261
0.4	-0.6	-0.91629073	-0.89995200	-0.01633873	01.78313839
0.3	-0.7	-1.20397280	-1.15297233	-0.05100047	04.23601512
0.2	-0.8	-1.60943791	-1.45860264	-0.15083525	09.37192077
0.1	-0.9	-2.30258509	-1.83012300	-0.47246209	20.51876799

Example 1-2   Now you try it. This time you will learn about the benefits of Chebyshev approximating polynomials as well as programming your 5100. Evaluate $\ln(1 + x)$ using the fifth-order Chebyshev approximating polynomial

$$\ln(1+x) \cong x(a_1 + x(a_2 + x(a_3 + x(a_4 + a_5 x)))), \qquad 0 \leqslant x \leqslant 1$$

over the range $0 \leqslant x \leqslant 1$ using the coefficients

$$a_1 = 0.99949556 \qquad a_4 = -0.13606275$$
$$a_2 = -0.49190896 \qquad a_5 = 0.03215845$$
$$a_3 = 0.28947478$$

If you have programmed your 5100 correctly, you will find the accuracy of this approximation as shown in Tables 1-4 and 1-5.

Note that even outside the region where the approximating polynomial was designed to best approximate $\ln(1 + x)$ it is more accurate than the "unconditioned" Taylor series expansion of $\ln(1 + x)$. Those who are interested should study Chebyshev economization techniques.

Examples 1-3 and 1-4 have nothing to do with programming your 5100 per se, but are important for learning how to develop algebraic expressions for use with the 5100. Sharpen your pencil and rearrange these expressions to minimize error in their function evaluation.

Table 1-4   Accuracy of the Fifth-Order Chebyshev Polynomial Approximation of ln(1 + x)

(1 + x)	x	ln(1 + x)	x[a₁ + x(a₂ + ···)]	Absolute Error	Relative Error (%)
1.1	+ 0.1	0.09531018	0.09530666	0.00000352	0.003697
1.2	+ 0.2	0.18232156	0.18233114	− 0.00000959	− 0.005257
1.3	+ 0.3	0.26236426	0.26236872	− 0.00000445	− 0.001697
1.4	+ 0.4	0.33647224	0.33646527	0.00000696	0.002070
1.5	+ 0.5	0.40546511	0.40545592	0.00000919	0.002267
2.0	+ 1.0	0.69314718	0.69315708	0.00000990	− 0.001428

Table 1-5   Accuracy of the Fifth-Order Chebyshev of ln(1 + x) Outside the Design Range of the Chebyshev Approximation

1 + x	x	ln(1 + x)	x[a₁ + x(a₂ + ···)]	Absolute Error	Relative Error (%)
0.9	− 0.1	− 0.10536052	− 0.10517205	− 0.00018847	0.178879
0.8	− 0.2	− 0.22314355	− 0.22211926	− 0.00102429	0.459028
0.7	− 0.3	− 0.35667494	− 0.35311655	− 0.00355840	0.997658
0.6	− 0.4	− 0.51082562	− 0.50084255	− 0.00998309	1.954301
0.5	− 0.5	− 0.69314718	− 0.66841824	− 0.02472894	3.567632
2.1	1.1	0.74193734	0.74210824	− 0.00017089	− 0.023034
2.2	1.2	0.78845736	0.78913899	− 0.00068162	− 0.086450
2.3	1.3	0.83290912	0.83478743	− 0.00187831	− 0.225512
2.4	1.4	0.87546874	0.87972822	− 0.00425948	− 0.486537
2.5	1.5	0.91629073	0.92481112	− 0.00852039	− 0.929878

Example 1-3   Rewrite the difference

$$h(x) = \frac{1}{x+1} - \frac{1}{x}$$

in a form that will minimize roundoff error using a series expansion technique. The objective is to eliminate the differencing of two numbers of approximately equal size. Expanding the first term, we see that

$$\frac{1}{x+1} = \frac{1/x}{1+1/x} = \frac{1}{x}\left(1 - \frac{1}{x} + \frac{1}{x^2} - \frac{1}{x^3} + \cdots\right). \qquad |x| > 1$$

Then

$$h(x) = \frac{1}{x}\left(1 - \frac{1}{x} + \frac{1}{x^2} - \frac{1}{x^3} + \cdots\right) - \frac{1}{x}$$

$$h(x) = \frac{1}{x}\left[\left(1 - \frac{1}{x} + \frac{1}{x^2} - \frac{1}{x^3} + \cdots\right) - 1\right]$$

$$h(x) = -\frac{1}{x^2}\left(1 - \frac{1}{x} + \frac{1}{x^2} - \cdots\right)$$

$$h(x) = \frac{-1}{x^2}\left(\frac{1}{1 + 1/x}\right) = \frac{-1}{x(x+1)}$$

This form of $h(x)$ does not involve computing the difference of two numbers of nearly equal size. The range of x over which this derivation applies is $|x| > 1$.

Example 1-4 Rewrite the difference

$$h(x) = \frac{1}{x+1} - \frac{1}{x}$$

in a form that will minimize roundoff error using algebra.
Cross-multiplying, we find

$$h(x) = \frac{x - (x+1)}{x(x+1)} = \frac{-1}{x(x+1)}$$

This result is the same as that developed with the series expansion method except that it holds for all x, not just $|x| > 1$. This is an important point to remember. Derivations using series expansion techniques *often* lead to results that hold over a greater range of the independent variable than their derivation strictly allows. With a pocket calculator it is easy to check the dynamic range over which a derived formula will work.

Example 1-5 Estimate $\sin(31°) - \sin(30°)$ using the mean value theorem. By the mean value theorem, we obtain

$$h(x) = \sin(30° + 1°) - \sin(30°) \approx 0.017453293 \cos(30.5°)$$

Here 0.017453293 is the value of 1° in radians. Then:

$$0.017453293 \cos(30.5°) = 0.015038266$$

$$\sin(31°) - \sin(30°) = 0.015038075$$

$$\text{relative error (\%)} = -0.0012700$$

$$\text{absolute error} = -0.000000191$$

Table 1-6 indicates that the mean value theorem can be useful for engineering evaluations, since the relative error is very small. Care must be taken, however, in using the mean value theorem. Had we used cos(30°) instead of cos(30.5°) we would find

$$0.017453293 \cos(30°) = 0.015114995$$

where actually

$$\sin(31°) - \sin(30°) = 0.015038075$$

$$\text{absolute error} = -0.000076920$$

$$\text{relative error (\%)} = -0.5115024$$

Table 1-6 Accuracy of Mean Value Theorem Approximation of $\sin(\theta + 1°) - \sin\theta$

θ (degrees)	Sin $(\theta + 1°) - \sin\theta$	Mean Value Theorem	Absolute Error	Relative Error (%)
0	0.017452406	0.017452628	−0.000000222	−0.0012
10	0.017160818	0.017161036	−0.000000218	−0.0012
20	0.016347806	0.016348014	−0.000000208	−0.0012
30	0.015038075	0.015038266	−0.000000191	−0.0012
40	0.013271419	0.013271588	−0.000000169	−0.0012
50	0.011101518	0.011101659	−0.000000141	−0.0012
60	0.008594304	0.008594412	−0.000000109	−0.0012
70	0.005825955	0.005824029	−0.000000074	−0.0012
80	0.002880587	0.002880624	−0.000000037	−0.0012
90	−0.000152305	−0.000152307	+0.000000002	−0.0012

Had we used cos(31°) instead of cos(30.5°), we would find

$$0.017453293 \cos(31°) = 0.014960392$$
$$\sin(31°) - \sin(30°) = 0.015038075$$
$$\text{absolute error} = 0.000077683$$
$$\text{relative error } (\%) = 0.5165751$$

Here we see that the relative error at the boundaries of the θ interval has jumped from $\sim\frac{1}{1000}\%$ when θ is taken at the midvalue of the interval to $\sim\frac{1}{2}\%$ when θ is taken at the end value of the interval.

Example 1-6 Rewrite the expression

$$f(A) = \frac{e^A \ln A}{e^{-A} + e^A} \qquad \text{(form 1)}$$

so that $f(A)$ can be evaluated at $A = 500$.

First note that if form 1 is programmed into the 5100 and evaluated, e^A will overflow at $A = 230.2585092$ and the expression cannot be evaluated. Rewriting to eliminate all terms of the form e^A, we find

$$f(A) - \frac{e^A \ln A}{e^{-A} + e^A}\left(\frac{e^{-A}}{e^{-A}}\right) = \frac{\ln A}{e^{-2A+1}} \qquad \text{(form 2)}$$

Program form 2 into the AER and you will find

$$f(500) = 6.214608098$$

Did you notice that for A on the order of 200, form 1 takes about three times as long to evaluate as form 2? The reason is that there are three terms of the form e^x to evaluate in form 1, whereas there is only one in form 2. To shorten the time for iterative problem solution, rewrite expressions to minimize redundant functions.

1-9 REFERENCE

For further efficiencies in algebraic function and infinite series evaluation, see Chebyshev economization techniqes in *Scientific Analysis on the Pocket Calculator*, J. M. Smith, 2nd ed., Wiley, New York (1977).

CHAPTER 2

DIFFERENCE TABLES AND DATA ANALYSIS

2-1 INTRODUCTION

This chapter deals with interpolation, extrapolation, and smoothing of tabulated data using numerical methods that are tailored to the Sharp 5100 scientific calculator. Many books on numerical analysis discuss these topics as related to the use of mathematical tables. Though we are interested in the use of these methods for precision table lookup, this chapter aims mainly to develop functions that are simple in form that can be used to replace complex functions. This technique, called analytic substitution, is commonplace in advanced analysis. For example, cost data developed on computer programs with as many as 500 cost-estimating relationships (CERs) can be used to generate a table of costs as a single design parameter is changed. It is often convenient to develop an inter-polation formula based on the table of discrete costs which will compute system cost as a function of the single design parameter. The simpler formula can be analytically substituted for the entire complex system of CERs in the large-scale cost model. This reduces the cost of "cost estimat-ing" and makes the simplified models convenient to analyze on the calculator. Finally, we will study what is perhaps the most important but seemingly least developed use of data tables, extrapolation or prediction. Here projections, predictions, and identification of trends and predicted values of function are discussed both from the viewpoints of mathematical limitations and the practical necessity to predict the behavior of dynamic processes from their data tables.

2-2 DIFFERENCE TABLES OF EQUALLY SPACED DATA

The difference tables that we are concerned with here are usually gener-ated in two ways. Either a function is evaluated for certain values of its independent variable or data are determined by measurements. In both cases the tables of equally spaced data can usually be prepared, especially of data determined from experiments, since much of experimental electronics and data sampling is done digitally and can be time-referenced to a digital clock. We discuss arbitrarily spaced data later in this chapter.

Our notation is based exclusively on the definition of the forward difference:

$$\Delta y_i = y_{i+1} - y_i = y(x_0 + [i+1]\Delta x) - y(x_0 + i\Delta x), \qquad i = 0, 1, 2, \ldots, n$$

Figure 2-1 illustrates the definitions of the differences involved in the difference table. Occasionally we use the term h to represent the spacing of the data, that is, $\Delta x = h = x_{n+1} - x_n$. We do not use backward differences or central differences in this book. Backward and central differences are only useful for changing the form of equations used in the derivation of numerical approximation methods. Since our interest here is not in manipulating equations but in their numerical evaluation, we use only the forward difference notation. Repeated application of the definition of the forward difference generates higher-order differences. For example, the

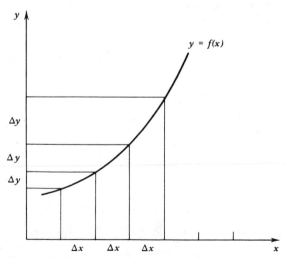

Figure 2-1   Definition of differences of equally spaced data.

second-order difference is derived as

$$\Delta^2 y_i = \Delta y_{i+1} - \Delta y_i$$

$$= y_{i+2} - y_{i+1} - (y_{i+1} - y_i)$$

$$= y_{i+2} - 2y_{i+1} + y_i$$

The third-order difference is developed as

$$\Delta^3 y_i = \Delta^2 y_{i+1} - \Delta^2 y_i$$

$$= \Delta y_{i+2} - \Delta y_{i+1} - (\Delta y_{i+1} - \Delta y_i)$$

$$= (y_{i+3} - y_{i+2}) - (y_{i+2} - y_{i+1}) - (y_{i+2} - y_{i+1}) + (y_{i+1} - y_i)$$

$$= y_{i+3} - 3y_{i+2} + 3y_{i+1} - y_i$$

The differences can be numerically evaluated using the equations just developed, or they can be computed directly from the tabulated values of the dependent variable, as shown in Figure 2-2.

The nth difference operator is given by the formula

$$\Delta^n = (z-1)^n = z^n - nz^{n-1} + \frac{n(n-1)}{2}z^{n-2} - \cdots \qquad (2\text{-}1)$$

where z is the shifting operator defined by the relation

$$z[y(x)] = y(x + \Delta x)$$

Furthermore, by repeated application of the shifting operator we see that

$$z^n[y(x)] = y(x + n\Delta x)$$

Equation 2-1 is derived by noting that the forward difference and shifting operators are related as follows:

$$\Delta y_i \quad = \quad y_{i+1} - y_i = zy_i - y_i = (z-1)y_i$$

$$\vdots \qquad \vdots \qquad\qquad \vdots$$

$$\Delta^n y_i \quad = \qquad\qquad (z-1)^n y_i$$

Note also that equation 2-1 can be written in the form

$$\Delta^n y_i = [z^n - C(n,1)z^{n-1} + C(n,2)z^{n-2} - \cdots] y_i$$

$$\Delta^n y_i = y_{i+n} - C(n,1)y_{i+n-1} + C(n,2)y_{i+n-2} - \cdots$$

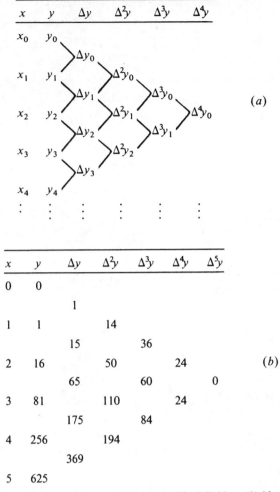

Figure 2-2 Finite difference tables. (*a*) Difference table definition. (*b*) Numerical example $y = x^4$.

where

$$C(n,r) = \frac{n!}{r!(n-r)!}$$

which is called the rth binomial coefficient of order n. The binomial coefficient is evaluated at a single key stroke on the 5100 using the combinations key nCr.

Example 2-1   Develop the table of finite differences shown in Figure 2-2.

Approach:

Step 1   Store the 6 values of y in the data storage registers A through F.

Step 2   Program the 5100 to take the finite difference and record the result in the second storage register used in the difference and display the result for tabulation.

Analysis:

  AER Mode

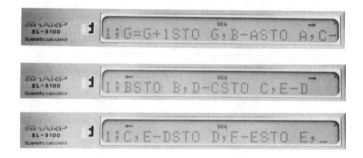

  COMP Mode

| 0 STO A | 1 STO B | 16 STO C | 81 STO D | 256 STO E |
| 625 STO F | 0 STO G | | | |

Now, by repeatedly keying *COMP* the following differences are developed.

The	first	stroke of	"*Comp*"	yields	Δy_1
The	second	"	"	"	"	Δy_2
The	third	"	"	"	"	Δy_3
The	fourth	"	"	"	"	Δy_4
The	fifth	"	"	"	"	Δy_5
The	sixth	"	"	"	"	a meaningless number
The	seventh	"	"	"	"	$\Delta^2 y_1$
The	eighth	"	"	"	"	$\Delta^2 y_2$
The	ninth	"	"	"	"	$\Delta^2 y_3$
The	tenth	"	"	"	"	$\Delta^2 y_4$
The	eleventh	"	"	"	"	a meaningless number
The	twelfth	"	"	"	"	a meaningless number

etc.

From this we can quickly prepare Table 2-1 of finite differences.

Table 2-1   Finite Differences

		Ans 1 indicates column to which the data applies				
x	y	Ans 1 Δy	Ans 1 $\Delta^2 y$	Ans 1 $\Delta^3 y$	Ans 1 $\Delta^4 y$	Ans 1 $\Delta^5 y$
0	0					
		Ans 2				
1	1		Ans 2			
		Ans 3		Ans 2		
2	16		Ans 3		Ans 2	
		Ans 4		Ans 3		Ans 2
3	81		Ans 4		Ans 3	
		Ans 5		Ans 4		
4	256		Ans 5			
		Ans 6				
5	625					

2-3 DATA INTERPOLATION

Armed with these definitions, we are now prepared to examine a number of formulas for analytic substitution or for interpolation. The method that we use here involves a Lozenge diagram of differences and binomial coefficients which can be combined into interpolation formulas. The diagram is shown in Figure 2-3. Certain rules applied along paths across the diagram proceeding from left to right define interpolation formulas. This diagram is so general that it encompasses both Newton's forward and backward difference formulas, Stirling's interpolation formula, Bessel's interpolation formula, and an interesting and unusual formula due to Gauss which zigzags across the diagram. The rules to be followed that generate these and many more interpolation formulas are the following:

1. When moving from left to right across the diagram, sum at each step.

2. When moving from right to left across the diagram, subtract at each step.

3. If the slope of the step is positive, the term in the interpolation formula for that step is the product of the difference crossed times the factor immediately below it.

4. If the slope of the step is negative, the term is the product of the difference crossed times the factor immediately above it.

5. If the step is horizontal and passes through a difference, the term is the product of the difference times the average of the factors above and below it.

6. If the step is horizontal and passes through a factor, the term is the product of the factor times the average of the differences above and below it.

Following these rules, starting at $y(0)$ and going down and to the right, we generate the interpolation formula

$$y(n) = y(0) + C(n,1)\Delta y(0) + C(n,2)\Delta^2 y(0) + \cdots$$

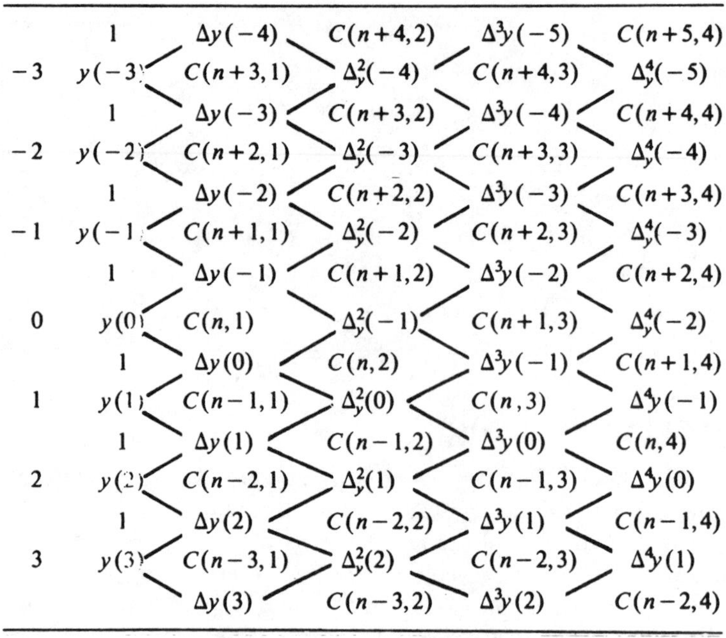

	1	$\Delta y(-4)$	$C(n+4,2)$	$\Delta^3 y(-5)$	$C(n+5,4)$
-3	$y(-3)$	$C(n+3,1)$	$\Delta_y^2(-4)$	$C(n+4,3)$	$\Delta_y^4(-5)$
	1	$\Delta y(-3)$	$C(n+3,2)$	$\Delta^3 y(-4)$	$C(n+4,4)$
-2	$y(-2)$	$C(n+2,1)$	$\Delta_y^2(-3)$	$C(n+3,3)$	$\Delta_y^4(-4)$
	1	$\Delta y(-2)$	$C(n+2,2)$	$\Delta^3 y(-3)$	$C(n+3,4)$
-1	$y(-1)$	$C(n+1,1)$	$\Delta_y^2(-2)$	$C(n+2,3)$	$\Delta_y^4(-3)$
	1	$\Delta y(-1)$	$C(n+1,2)$	$\Delta^3 y(-2)$	$C(n+2,4)$
0	$y(0)$	$C(n,1)$	$\Delta_y^2(-1)$	$C(n+1,3)$	$\Delta_y^4(-2)$
	1	$\Delta y(0)$	$C(n,2)$	$\Delta^3 y(-1)$	$C(n+1,4)$
1	$y(1)$	$C(n-1,1)$	$\Delta_y^2(0)$	$C(n,3)$	$\Delta^4 y(-1)$
	1	$\Delta y(1)$	$C(n-1,2)$	$\Delta^3 y(0)$	$C(n,4)$
2	$y(2)$	$C(n-2,1)$	$\Delta_y^2(1)$	$C(n-1,3)$	$\Delta^4 y(0)$
	1	$\Delta y(2)$	$C(n-2,2)$	$\Delta^3 y(1)$	$C(n-1,4)$
3	$y(3)$	$C(n-3,1)$	$\Delta_y^2(2)$	$C(n-2,3)$	$\Delta^4 y(1)$
		$\Delta y(3)$	$C(n-3,2)$	$\Delta^3 y(2)$	$C(n-2,4)$

Figure 2-3 The Lozenge diagram.

which becomes

$$y(n) = y(0) + n\Delta y(0) + \frac{n(n-1)}{2}\Delta^2 y(0) + \cdots$$

This is Newton's forward difference interpolation formula. To generate Newton's backward difference formula, the procedure is reversed. Starting at $y(0)$ and moving up and to the right, we generate the formula

$$y(n) = y(0) + C(n,1)\Delta y(-1) + C(n+1,2)\Delta^2 y(-2) + \cdots$$

which becomes

$$y(n) = y(0) + n\Delta y(-1) + \frac{n(n+1)}{2}\Delta^2 y(-2) + \cdots$$

This is Newton's backward difference formula.

To develop Stirling's formula, we start at $y(0)$ and move horizontally to the right. In this case, we generate the interpolation formula

$$y(n) = y(0) + C(n,1)\left\{ \frac{\Delta y(0) + \Delta y(-1)}{2} \right\}$$

$$+ \left\{ \frac{C(n+1,2) + C(n,2)}{2} \right\} \Delta^2 y(-1) + \cdots$$

$$y(n) = y(0) + n\left\{ \frac{\Delta y(0) + \Delta y(-1)}{2} \right\} + \frac{n^2}{2}\Delta^2 y(-1) + \cdots$$

Bessel's formula can be generated by starting midway between $y(0)$ and $y(1)$.

$$y(n) = 1\left\{ \frac{y(0) + y(1)}{2} \right\} + \left\{ \frac{C(n,1) + C(n-1,1)}{2} \right\} \Delta y(0) + \cdots$$

$$y(n) = \left\{ \frac{y(0) + y(1)}{2} \right\} + (n - \tfrac{1}{2})\Delta y(0)$$

$$+ \frac{n(n-1)}{2}\left\{ \frac{\Delta^2 y(-1) + \Delta^2 y(0)}{2} \right\} + \cdots$$

A great number of other formulas can be generated and used for interpolation of data.

Interpolation is often employed in computing intermediate values of tabulated functions. The Sharp 5100 scientific calculator gives sine, cosine, tangent, arc sine, arc cosine, arc tangent, hyperbolic sine, hyperbolic cosine, and hyperbolic tangent, so there is no need for interpolation of data tables for these functions. Interpolation is important when we must use tables of Bessel functions, Legendre polynomials, error functions, and the like or when using tables of experimental data. Those are often more easily evaluated with standard reference tables. In these cases, it is occasionally necessary to interpolate between two values in the table.

Before discussing the interpolation process, however, it is worth pointing out that most well-made tables are often generated with auxiliary functions as opposed to the actual functions themselves. For example, the exponential integral with positive argument is given by

$$Ei(x) = \int_{-\infty}^{x} \frac{e^u}{u}\,du$$

which takes the series form

$$Ei(x) = \gamma + \ln(x) + \frac{x}{1 \cdot 1!} + \frac{x^2}{2 \cdot 2!} + \frac{x^3}{3 \cdot 3!} + \cdots$$

can be approximated with the series

$$Ei(x) \cong \frac{e^x}{x}\left[1 + \frac{1!}{x} + \frac{2!}{x^2} + \frac{3!}{x^3} + \cdots\right] \qquad (x \to \infty)$$

The logarithmic singularity in the first series does not permit easy interpolation near $x = 0$. The function $Ei(x) - \ln x$ is better behaved and more readily interpolated when x is near zero. In fact, $x^{-1}[Ei(x) - \ln(x) - \gamma]$ (where γ is Euler's constant $0.577\cdots$) is an auxiliary function that results in a slightly higher interpolation accuracy than when $Ei(x)$ is computed from interpolated values of the table of $Ei(x)$ directly.

Generally tables are constructed and presented so that reasonable-order interpolating polynomials (i.e., first-, second-, or third-order) can be used to compute intermediate values while retaining the precision of the table.

For example, in the *Handbook of Mathematical Functions* (U.S. Department of Commerce, Bureau of Standards, Applied Mathematics Series 55) most tables are accompanied by a statement of the maximum error in a linear interpolation between any two numbers in the table, and the number of function values needed in Laplace's formula or Atkins' method to interpolate to nearly full tabular accuracy.

An example from the *Handbook of Mathematical Functions* appears in Table 2-2. The accuracy statement is given in brackets. The numbers in brackets mean that the maximum error in a linear interpolate is 3×10^{-6} and that to interpolate to the full tabular accuracy, five points must be used in Lagrange's method or Atkins' method of interpolation. The linear interpolation formula is

$$f_p = (1 - p)f_0 + pf_1$$

which when programmed on the 5100 in the AER mode takes the form

Table 2-2 Exponential Integral Auxiliary Function

x	$xe^{x}E_1(x)$	x	$xe^{x}E_1(x)$
7.5	0.892687854	8.0	0.898237113
7.6	0.893846312	8.1	0.899277888
7.7	0.894979666	8.2	0.900297306
7.8	0.896088737	8.3	0.901296023
7.9	0.897174302	8.4	0.902274695

$$\begin{bmatrix} (-6)3 \\ 5 \end{bmatrix}$$

Here p is stored in A, f_0 is stored in B, and f_1 is stored in C. Here f_0, f_1 are consecutive tabular values of the function corresponding to arguments x_0, x_1 respectively; p is the given fraction of the argument interval

$$p = \frac{(x - x_0)}{(x_1 - x_0)}$$

and f_p is the required interpolate. For example, if we interpolate between the values of Table 2-2 for $x = 7.9527$, we find that

$$f_0 = 0.897174302; \quad \text{STO B}$$

$$f_1 = 0.898237113; \quad \text{STO C}$$

$$p = 0.527; \quad \text{STO A}$$

from which we obtain

$$f_{0.527} = 0.897734403.$$

The terms in the brackets of Table 2-2 indicate that the accuracy for linear interpolation is 3×10^{-6}. Thus we round this result to 0.89773 by setting the display to TAB-5. The maximum possible error in this answer is composed of the error committed by the last rounding, that is, $0.4403 \times 10^{-5} + 3 \times 10^{-6}$, and thus cannot exceed 0.8×10^{-5}.

To get greater precision, we can interpolate this example of the table using Lagrange's formula. In this example, the interpolation formula is the five-point formula:

$$f(x_0 + p\Delta x) = \left\{ \frac{(p^2-1)(p-2)p}{24} \right\} f_{-2} - \left\{ \frac{(p-1)(p^2-4)p}{6} \right\} f_{-1}$$

$$+ \left\{ \frac{(p^2-1)(p-2)p}{4} \right\} f_0 - \left\{ \frac{(p+1)(p^2-4)p}{6} \right\} f_1$$

$$+ \left\{ \frac{(p^2-1)(p+2)p}{24} \right\} f_2, \qquad |p| < 1$$

Another approach is to use a five-term Newton forward or backward difference formula, a Bessel's formula, Stirling's formula, or any of the formulas that come out of the Lozenge diagram. The details associated with such interpolations are conveniently found in Chapter 25 of the *Handbook of Mathematical Functions*.

Since there are occasions for using inverse interpolation, we discuss it briefly here. If we are given a table of values of the dependent variable y_n as a function of values of the independent variable x_n,

$$y_n = f(x_n) \qquad \text{(tabulated function)}$$

then intermediate values of y can be computed by interpolating between the values y_n with an interpolating polynomial $g(x)$ as

$$y = g(x) \cong f(x) \qquad \text{(continuous function)}$$

Inverse interpolation is a matter of viewpoint. Here we would view the interpolation from the standpoint of the dependent variable,

$$x_n = f^{-1}(y_n) \qquad \text{(tabulated function)}$$

Then intermediate values of x can be computed by interpolating between the values of x_n with an interpolating polynomial $h(y)$ as

$$x = h(y) \cong f^{-1}(y) \qquad \text{(continuous function)}$$

With linear interpolation there is no difference in principle between direct and inverse interpolation. In cases where the linear formula is not sufficiently accurate, two methods are available for accuracy improvement. The first is to interpolate more accurately by using, for example, a higher-order Lagrange's formula or an equivalent higher-order polynomial method. The second is to prepare a new table with a smaller interval in the neighborhood of interest, and then apply accurate inverse linear interpolation to the subtabulated values.

It is important to realize that the accuracy of inverse interpolation may be very different from that of a direct interpolation. This is particularly true in regions where the function is slowly varying, such as near flat maximum or minimum. The maximum absolute error resulting from inverse interpolation can be estimated with the aid of the formula

$$\delta x = \left(\frac{\partial f}{\partial x} \right)^{-1} \delta y, \quad \delta x \approx \left(\frac{\Delta f}{\Delta x} \right)^{-1} \delta y$$

where δy is the maximum possible error in the tabulation of y values and Δf and Δx are the first differences generated from the table in the neighborhood of the region of interest.

Let us now return our attention to the generation of difference tables. In the generation of interpolating polynomials of reasonable size, the finite differences in the difference table must be small for high-order differences. If they are not small the questions is, "What can we do to reduce the size of the finite differences?"

There are only three considerations associated with any difference table. The first is the number of differences to which the table is taken the second is the spacing between different values of the tabulated function, and the third is the number of figures tabulated. The effect of halving (factor of $\frac{1}{2}$) spacing in the independent variable x is to divide the first differences by 2, the second differences by 4, the third differences by 8, and so on. Examples of the effect that different spacings of x have on the function $y = x^3$ appear in Table 2-3. In answer to the question above then: to reduce the nth-order difference by a factor k we must reduce the data interval (independent variable) by a factor of approximately $\sqrt[n]{k}$.

2-4 DATA EXTRAPOLATION

Extrapolation outside the range of data that makes up a difference table is a controversial procedure. It is, however, a procedure of great practical interest. Given the behavior of a dynamic process, sampled at intervals, it is only natural to ask to what extent the table can be extended beyond the range of the data used to make up the table to *predict* the future behavior of the process being considered. This is a very practical, important, and real matter. It is the problem of science to be interested in predicting the behavior of systems based on observations of their past behavior. While there are a number of stock market "chartsmen" who use finite difference techniques, it is generally accepted that extrapolation outside the range of the difference table is as much an art as a science. Because of its practical

Table 2-3 The Effect of Interval Halving on The Finite Differences in a Difference Table[a]

Full interval $\Delta x = 2$					Half-interval $\Delta x = 1$				
x	$y = x^3$	Δy	$\Delta^2 y$	$\Delta^3 y$	x	$y = x^3$	Δy	$\Delta^2 y$	$\Delta^3 y$
0	0				0	0			
		8					1		
2	8		48		1	1		6	
		56		48			7		6
4	64		96		2	8		12	
		152		48			19		6
6	216		144		3	27		18	
		296		48			37		6
8	512		192		4	64		24	
		488					61		
10	1000				5	125			

[a]Third-order difference is reduced by a factor of 8 when interval between values of x is halved.

value and practical interest, it will be covered here but with the proviso that the reader recognize that extrapolation is, in theory, a questionable procedure. That is, the same difference table using only slightly different extrapolation techniques can, and usually does, lead to significantly different predictions. Because of this lack of robustness of extrapolated data, the procedure has questionable value.

We illustrate the problem of prediction with the following practical example. Consider an aircraft executing a fully automatic landing. Sampled values of the altitude are shown in Table 2-4. What will be the conditions at touchdown? This particular example is a nontrivial one, in that the heart of present-day flight-control performance monitors hinges on the ability to predict the dynamic behavior of high-energy devices, such as aircraft, when terminal operations are under automatic control. The obvious first step is to form the difference table, as shown in Table 2-4. We note that this table can be carried to the third difference without the lower-order differences becoming constant. The obvious next step to extrapolation is to assume that the third difference holds constant up to touchdown, and to predict the behavior shown in Table 2-5. Clearly the predicted results are fairly grim. We have low confidence in extrapolations of this type because the difference table did not indicate the influence of any control law through arriving at constant differences between any of the finite differences. Had we found, for example, that all of the second differences held constant and the third differences were zero, we might be

Table 2-4 Difference Table of Altitude of an Aircraft Executing an Automatic Landing

t	$h(t)$	Δh	$\Delta^2 h$	$\Delta^3 h$
0	60			
		-13		
1	47		$+3$	
		-10		-2
2	37		$+1$	
		-9		0
3	28		$+1$	
		-8		
4	20			

entitled to higher confidence in the extrapolation to touchdown by assuming that the guidance law objective was to hold the third-order differences to zero. It follows, then, that a procedure for increasing the confidence in extrapolation from finite difference tables is to find a transformation of the variable of interest which would expose the guidance law and its effect on the difference table.

Table 2-5 Extrapolated Touchdown Conditions of Automatically Landed Aircraft

	t	$h(t)$	Δh	$\Delta^2 h$	$\Delta^3 h$	
Range of actual data	0	60				
			-13			
	1	47		$+3$		
			-10		-2	Average $=-1$
	2	37		$+1$		
			-9		0	
	3	28		$+1$		
			-8		-1	
	4	20		0		
			-8		-1	
Range of extrapolation	5	12		-1		
			-9		-1	
	6	Touchdown sink rate 3		-2		
			-11			
	7	~11 fps[a] (hard) -8				

[a] fps = feet per second.

After a little thought, it might be expected that the guidance law to automatically land an aircraft would be an exponential law of the form

$$\frac{dh}{dt} = -k(h + h_B)$$

which results in a flared landing path of the form

$$h = h_0 e^{-kt} - h_B$$

which is sketched in Figure 2-4. This suggests the formation of the difference Table 2-6. Note that the second difference is near zero so that for short-term prediction (next 7 seconds) a reasonable assumption is that $\Delta \ln(h)$ is approximately constant and equal to -0.3062. The differences between the two approaches are tabulated in Table 2-7. It is apparent that the logarithmic extrapolation does better short-term prediction than does the "third-order" extrapolation.

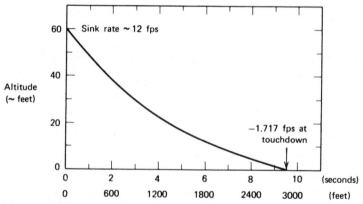

Figure 2-4   Typical jet transport landing trajectory (fps = feet per second).

In summary, we might expect to extrapolate with greater confidence from Table 2-6 than Table 2-5 because of the observed characteristics of the guidance law. From a strictly mathematical viewpoint, the issue is not all that clear. The number of actual samples of the second difference is small and thus the true mathematical confidence in the fact that the guidance law is in some way holding the second difference to zero is low. In the end, extrapolation using difference tables involves careful judgment.

Table 2-6 Logarithmic Extrapolation of Touchdown Conditions of Automatically Landed Aircraft

	t	$h(t)$	$\ln(h(t))$	$\Delta\ln(h(t))$	
	0	60	4.0943		
				− 0.2442	
	1	47	3.8501		
Actual				− 0.2392	
data	2	37	3.6109		Average ≅ − 0.27465
range				− 0.2787	
	3	28	3.3322		
				− 0.3365	
	4	20	2.9957		
				− 0.27465	
	5	15.2	2.7405		
				− 0.27465	
	6	11.4	2.4464		
				− 0.27465	
	7	8.7	2.17175		
Range				− 0.27465	
of ex-	8	6.7	1.89710		
trapo-				− 0.27465	
lation	9 Sink	5.1	1.62245		
	rate			− 0.27465	
	10 ∼ 1.3lfps[a]	3.8	1.3478		
				− 0.27465	
	11	2.9	1.07315		

[a] fps = feet per second.

Table 2-7 Comparison of Extrapolation Methods for Predicting Touchdown Conditions

	Actual	"Third-Order" Extrapolation		First-Order Logarithmic Extrapolation	
t	h	h	Absolute Error	h	Absolute Error
5	16	12	−4	15	−1
6	11	3	−8	11	0
7	7	−8	−15	9	+2
8	4			7	+3
9	1.6			5	+3.4
10	0			4	+4
11	0			3	+3

[a] fps = feet per second.

2-5 DATA ERROR LOCATION AND CORRECTION

Errors due to observations, calculation, measurement, or recording often occur in a table of numbers. These errors introduced into the calculation process are significantly magnified in the generation of ascending differences in the difference table. This can be seen in Table 2-8. It is apparent that the errors propagate and are distributed binomially (in any given difference the errors are weighted by binomial coefficients). It is also apparent that the error grows rapidly as it propagates into ascending orders of difference. For example, the error in Table 2-9 might be anticipated by noting the form of the third difference. We see the pattern of signs $(+)$, $(-)$, $(+)$, $(-)$ indicative of error propagation. Also, note that the pattern of fourth difference is centered on $y = 17$. Furthermore, note

Table 2-8   Error Propagation in Difference Tables

y	Δy	$\Delta^2 y$	$\Delta^3 y$	$\Delta^4 y$
0				
	0			
0		0		
	0		0	
0		0		$+\epsilon$
	0		$+\epsilon$	
0		$+\epsilon$		-4ϵ
	$+\epsilon$		-3ϵ	
$+\epsilon$		-2ϵ		$+6\epsilon$
	$-\epsilon$		$+3\epsilon$	
0		$+\epsilon$		-4ϵ
	0		$-\epsilon$	
0		0		$+\epsilon$
	0		0	
0		0		
	0			
0				

Table 2-9   Unit Error Propagation in the Difference Table for the Function $y = x^2$

x	y	Δy	$\Delta^2 y$	$\Delta^3 y$	$\Delta^4 y$
0	0				
		1			
1	1		2		
		3		0	
2	4		2		-1
		5		-1	
3	9		3		-4
		8		-3	
4	17		0		+6
		8		+3	
5	25		3		-4
		11		-1	
6	36		2		+1
		13		0	
7	49		2		
		15			
8	64				

that 6ϵ in Table 2-8 corresponds to 6 in Table 2-9; that is,

$$6\epsilon = 6$$

$$\epsilon = 1$$

Moreover, if the error in the values of y were of the form

$$y = x^2 + 5$$

we can expect in the fourth difference column to show an error of $6k$. Thus one-sixth of the fourth difference which is centered on the number in error is a measure of the error—which can then be subtracted from the column of y values. We might modify Table 2-9 by replacing 17 with $(17-1) = 16$, thus obtaining the difference table shown in Table 2-10. In general, then, data smoothing is done by:

1. Keeping an eye open for the $(+)$, $(-)$, $(+)$, $(-)$,... pattern in high-order differences that indicates error propagation.
2. Identifying the tabulated value on which the pattern is centered.

Table 2-10 Smoothed Data Table for the Function
$y = x^2$

x	y	Δy	$\Delta^2 y$	$\Delta^3 y$	$\Delta^4 y$
0	0				
		1			
1	1		2		
		3		0	
2	4		2		0
		5		0	
3	9		2		0
		7		0	
4	16		2		
		9			
5	25				

3. Equating observed error with its binomial error counterpart.
4. Solving for the error and appropriately modifying the data table.
5. Testing the table for elimination of the $(+), (-), (+), (-), \ldots$ pattern.

2-6 MISSING ENTRIES

Occasionally a difference table has a few missing entries in the dependent variable. Missing entries in the difference table can be estimated in several ways.

The simplest method is to examine the table and decide whether the points could be reasonably fit with a polynomial. For example, a data table with four points, one of which is unknown, might be fit with a second-degree polynomial. It is characteristic of difference tables that nth-order differences of polynomials of degree $n-1$ equal zero. For example, the equation

$$y = 2x^2 + x + 3$$

has the difference equation shown in Table 2-11, where it is apparent that the third-order differences equal zero. This characteristic is present in general in nth-order polynomials; that is, their $(n+1)$st-order (and all higher) differences equal zero. Using this property, we would expect that the fourth-order difference would equal zero; that is

$$\Delta^4(y) = 0$$

Table 2-11   Difference Table for $y = 2x^2 + x + 3$

Subscript in Missing Entry Formula	x	y	Δy	$\Delta^2 y$	$\Delta^3 y$
	0	3			
			3		
	1	6		4	
			7		0
0	2	13		4	
			11		0
1	3	24		4	
			15		0
2	4	39		4	
			19		0
3	5	58		4	
			23		0
4	6	81		4	
			27		
	7	108			

This can be rewritten in the shifting-operator notation as

$$(z-1)^4 y = (z^4 - 4z^3 + 6z^2 - 4z + 1)y = 0$$

This gives us

$$y_4 - 4y_3 + 6y_2 - 4y_1 + y_0 = 0$$

We can now solve for y^2 as

$$1; \quad f(ABCD) = (4(B + C) - (4 + D)) \div 6$$

where

$$A = y_0$$
$$B = y_1$$
$$C = y_3$$
$$D = y_4$$

Here we use an even-order difference because all even-order difference equations do the following:

1. Give one middle term, which can be centered in the missing number in the table.

2. Result in missing entry determination with a minimum of roundoff error.

3. Are numerically more stable than their odd-order counterparts.

Let us assume, for the sake of the discussion, that the $y = 39$ entry is missing in Table 2-11. We can substitute directly from the table with the missing data point to obtain

$$y_2 = \tfrac{1}{6}[4(24 + 58) - (81 + 13)]$$

from which we can solve for the missing data point:

$$y_2 = \tfrac{1}{6}[4(82) - 94] = \tfrac{1}{6}(328 - 94 = 234) = \tfrac{234}{6} = 39$$

The method just described for filling in missing values in the data table is particularly suited to analysis on the 5100 in that it does not involve the determination of unknown coefficients in a polynomial (the usual methods for missing data determination). For tables with large numbers, the arithmetic could be tedious, but with the 5100 it is a simple matter to perform the sums and products for tables of large values requiring high precision. Another point worth making regarding identification of missing entries in data tables is that for tables with large numbers of values, say on the order of 20 to 100, it is not necessary to look for twentieth-order differences to develop the formula for computing the missing data. One need only determine the polynomial that can be reasonably expected to fit locally through two, four, or six data points symmetrically placed about the missing value to find the missing point.

We have been stressing the determination of interpolating polynomials by way of finite difference tables because the 5100 enables one to find finite differences quickly and conveniently, thus leading immediately to interpolation formulas of high order and high accuracy, which themselves can be evaluated on the 5100 conveniently and to high precision. This, in fact, is the reason for using the calculator with difference tables: high-order difference tables lead to high-order approximating polynomials, which, when written in nested parenthetical form, are easily evaluated on the calculator to high precision.

The difficulty in using low-order polynomials for manual analysis in the precalculator era was that they generally were not sufficiently accurate to permit the precision numerical evaluation necessary for most engineering, economic, chemical, and other types of precision analysis. With the Sharp 5100 scientific calculator we can conduct precision analysis relatively quickly and efficiently by using high-order polynomials generated simply with difference tables of high order.

2-7 LAGRANGE'S INTERPOLATION FORMULA

So far we have studied the interpolation of equally spaced data through the use of difference tables and the Lozenge diagram as a convenient means for remembering a large number of different interpolation formulas. These interpolation formulas, however, do not apply to nonequally spaced values of the independent variable nor when the nth differences of the dependent variable are not small or zero. In these cases we can use Lagrange's interpolation formula to develop a polynomial that can be used for analytic substitution. Though there are other interpolation formulas for unequally spaced data, the advantage to using Lagrange's interpolation formula is that the coefficients are particularly easy to remember, and to determine, with the calculator. The method works for both nonequally and equally spaced data and regardless of whether the nth differences are small. Lagrange's interpolation formula is

$$y = y_0 \frac{(x - x_1)(x - x_2) \cdots (x - x_p)}{(x_0 - x_1)(x_0 - x_2) \cdots (x_0 - x_p)} + y_1 \frac{(x - x_0)(x - x_2) \cdots (x - x_p)}{(x_1 - x_0)(x_1 - x_2) \cdots (x_1 - x_p)}$$

$$+ \cdots + y_p \frac{(x - x_0)(x - x_1) \cdots (x - x_{p-1})}{(x_p - x_0)(x_p - x_1) \cdots (x_p - x_{p-1})}$$

An interesting and important feature of Lagrange's interpolation formula is that, if the data table has n entries, the formula appears to have n terms. It turns out, however, that if the table amounts to four or five samples of, say, a second-order polynomial, the terms will cancel, giving only the pieces due to the quadratic function. Because of roundoff the exact cancellation of the coefficients for the higher powers of x will not occur, but they will be very small, indicating that they should be made zero.

CHAPTER 3

ELEMENTARY FORMULAS FOR THE 5100 SCIENTIFIC CALCULATOR

3-1 INTRODUCTION

This chapter does not deal with programming the 5100 to solve problems. This chapter contains many useful formulas that occur in elementary analysis that are sized and in a form for immediate use on the 5100. They are presented here for the sake of convenience and completeness in a book on numerical analysis with the 5100. How these formulas are programmed on the 5100 is of much less importance than knowing how to use the formulas correctly. So the emphasis here is on understanding the mathematics more than the 5100 per se.

3-2 NUMERICAL EVALUATION OF PROGRESSIONS

An arithmetic progression is defined by a sequence of numbers

$$a_n = a_1 + (n-1)d, \quad (n \text{ an integer } > 0)$$

where a and d are real numbers. For $a_1 = 3e$ and $d = -\pi$

n	a_n
1	8.154845484
2	5.013252830
3	1.871660176
4	-1.269932478
5	-4.411525132

A common problem is to compute the sum of the arithmetic progression to n terms:

$$S_n(d) = a + (a+d) + (a+2d) + \cdots + [a+(n-1)d]$$

There are two formulas for computing the sum of an arithmetic progression. The first is

$$S_n(d) = na + \tfrac{1}{2}n(n-1)d$$

which can be rewritten in nested parenthetical form for efficient evaluation on the 5100 as

Here

$$f(ABC) = S_n(d)$$
$$A = n$$
$$B = a$$
$$C = d$$

Another formula for computing the sum of the arithmetic progression to n terms is

$$S_n(d) = \frac{n}{2}(a+l)$$

Here the last term in the series l is

$$l = a + (n-1)d$$

We note that this equation is already in a form that can be easily evaluated on the pocket calculator.

The geometric progression is defined by a sequence of terms of the form

$$a_n = a_1 r^{n-1}, \qquad (n \text{ an integer } > 0)$$

where a and r are real numbers. For $a_1 = 3e$ and $r = -\pi$

n	a_n
1	8.154845484
2	-2.561920267×10^1
3	$+8.048509891 \times 10^1$
4	-2.528513955×10^2
5	$+7.943560867 \times 10^2$

The sum of the geometric progression to n terms is

$$S_n = a_1 + a_1 r + a_1 r^2 + a_1 r^3 + a_1 r^4 + \cdots + a_1 r^{n-1}$$

It can be computed with the formula

$$S_n = \frac{a_1(1 - r^n)}{1 - r} = \frac{a_1 - rl}{1 - r}$$

where l is the last term. S_n can be evaluated on the 5100 with the expression

$$1; \quad \text{f(ABC)} = \text{A} \times (1 - \text{BY}^\text{x}\text{C}) \div (1 - \text{B})$$

where

$$\text{f(ABC)} = S_n$$
$$\text{A} = a_1$$
$$\text{B} = r$$
$$\text{C} = n$$

If $r < l$ in size, then as $n \to \infty$

$$\lim_{n \to \infty} (S_n) = \frac{a_1}{1 - r}$$

since the last term $l \to 0$. The sum of the geometric progression to n terms requires scratch-pad or memory storage. Table 3-1 shows a typical key stroke sequence needed for its evaluation and the required storage.

Three types of means are encountered in advanced analysis—the arithmetic mean, the geometric mean, and the harmonic mean. Though they are all special cases of the generalized mean

$$M(t) = \left(\frac{1}{n} \sum_{k=1}^{n} a_k^t \right)^{1/t}$$

we are explicit here and write them out. The arithmetic mean of n quantities is defined by the equation

$$A_n = \frac{a_1 + a_2 + a_3 + \cdots + a_n}{n}$$

which can be computed conveniently using a recursion formula

$$A_{n+1} = \frac{1}{n+1}(nA_n + a_{n+1})$$

An advantage to using a recursive "averager" is that the analyst can observe the convergence of the mean as he adds more terms to the calculation. He can thus often reduce the workload in computing an average by using only the numbers that are necessary to estimate the mean to the accuracy he desires.

The recursive form is directly implementable in the AER mode using the function

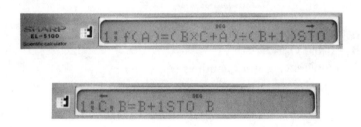

and

$$B_0 = C_0 = 0$$

Here

$$A = a_{n+1}$$
$$B = n$$
$$C = A_{n+1}$$

The geometric mean of n quantities is defined by the relationship

$$G = (a_1 a_2 \cdots a_n)^{1/n}, \qquad (a_i > 0, i = 1, 2, \ldots, n)$$

which is easily calculated using the recursion formula for the geometric mean of n quantities as given by

$$G_{n+1} = (a_{n+1} G_n^n)^{1/n+1}$$

This recursive form can be implemented on the 5100 in the AER mode using the expression

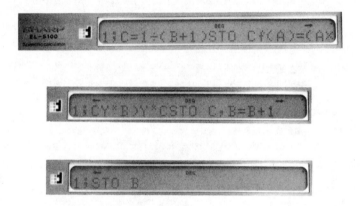

where $B_0 = 0$ and $G_0 = 1$.

The harmonic mean of n quantities is defined by

$$\frac{1}{H} = \frac{1}{n}\left(\frac{1}{a_1} + \frac{1}{a_2} + \cdots + \frac{1}{a_n}\right), \qquad (a_i > 0, i = 1, 2, \ldots, n)$$

It, too, can be evaluated by using a recursion formula:

$$H_{n+1} = \left\{\frac{1}{n+1}\left(\frac{1}{a_{n+1}} + \frac{n}{H_n}\right)\right\}^{-1}$$

Finally, the generalized mean is related to the geometric, arithmetic, and harmonic means according to the relations

$$\lim_{t \to 0} M(t) = G$$

$$M(1) = A$$

$$M(-1) = H$$

3-3 THE DEFINITION OF ABSOLUTE AND RELATIVE ERROR

We discussed absolute and relative errors previously in a general way in the context of other matters. It is important to define precisely what is meant by absolute and relative error. When x_0 is an approximation to the true value of x, we say the following:

1. The absolute error of x_0 is $\Delta x = x_0 - x = $ (calculated $-$ true).
2. The relative error of x_0 is $\delta x = \Delta x / x$, (calculated $-$ true)/true, which is approximately equal to $\Delta x / x_0$.
3. The percentage error is 100 times the relative error.

If in (2) we use the approximation of the true value of x to estimate percentage error then in a sense there is a small error in estimating the relative error.

The absolute error of the sum or difference of several numbers is *at most* equal to the sum of the absolute errors of the individual numbers. If it can be assumed that the errors occur in a random independent fashion, a more reasonable estimate of the error in computing the sum or difference of several numbers is root-sum-square error defined as

$$\left(\sum \Delta x_i^2 \right)^{1/2}$$

The relative error of the product or quotient of several factors is at most equal to the sum of the relative errors of the individual factors. Finally, if $y = f(x)$, the relative error

$$\delta y = \frac{\Delta y}{y} \cong \frac{f'(x)}{f(x)} \Delta x$$

If we have

$$y = f(x_1, x_2, \ldots, x_n)$$

and the absolute error in x_i is Δx_i for all n, then the absolute error in f is

$$\Delta f \approx \frac{\partial f}{\partial x_1} \Delta x_1 + \frac{\partial f}{\partial x_2} \Delta x_2 + \cdots + \frac{\partial f}{\partial x_n} \Delta x_n$$

Simple rules, similar to those for the relative error of a product or the quotient, can easily be derived for relative errors of powers and roots. It turns out that the relative error of an nth power is almost exactly n times the relative error of the base power, while the relative error of an nth root is $1/n$th of the relative error of the radicand.

3-4 INFINITE SERIES

Taylor's formula for approximating a single variable function is given by the expression

$$f(x+h) = f(x) + hf'(x) + \frac{h^2}{2}f''(x) + \cdots + \frac{h^{n-1}}{(n-1)!}f^{n-1} + Rn$$

This equation has an error formula that can be written in three typical forms:

$$R_n = \frac{h^n}{n!}f^n(x+\theta_1 h), \qquad (0 < \theta_1 < 1)$$

$$R_n = \frac{h^n}{(n-1)!}(1-\theta_2)^{n-1}f^n(x+\theta_2 h), \qquad (0 < \theta_2 < 1)$$

$$R_n = \frac{h^n}{(n-1)!}\int_0^1 (1-t)^{n-1}f^n(x+th)dt$$

The truncated version of the series can be expanded in nested parenthetical forms for efficient numerical evaluations when the numerical values of the derivative either are given or can be quickly computed.

$$f_1 = f(x)$$

$$f_2 = f(x) + hf'(x)$$

$$f_3 = f(x) + h\left(f' + \frac{hf''}{2}\right)$$

$$f_4 = f(x) + h\left(f' + \frac{h}{2}\left(f'' + \frac{hf'''}{3}\right)\right)$$

$$f_5 = f(x) + h\left(f' + \frac{h}{2}\left(f'' + \frac{h}{3}\left(f''' + \frac{hf''''}{4}\right)\right)\right)$$

$$\vdots$$

$$f_n = f(x) + h\left(f' + \frac{h}{2}\left(f'' + \frac{h}{3}\left(f''' + \frac{h}{4}\left(f'''' + \cdots \right. \right.\right.\right.$$

$$\left. \left. \left. \left. + \frac{h}{n-1}\left(f^{n-1} + \frac{hf^n}{n}\right)\right) \cdots \right)\right)\right)$$

Taylor series expansions of $f(x)$ on the point a are given by the expression

$$f(x) = f(a) + (x-a)f'(a) + \frac{(x-a)^2}{2}f''(a) + \cdots + \frac{(x-a)^{n-1}}{(n-1)!}f^{n-1}(a) + R_n$$

where the remainder formula is given by

$$R_n = \frac{(x-a)^n}{n!}f^n(\zeta), \qquad (a < \zeta < x)$$

This expression, too, can be written in nested parenthetical form as

$$f_n = f(a) + (x-a)\left(f'(a) + \frac{(x-a)}{2}\left(f''(a) + \frac{(x-a)}{3}\left(f'''(a) + \cdots \right.\right.\right.$$

$$\left.\left.\left. + \frac{(x-a)}{n-1}\left(f^{n-1}(a) + \frac{(x-a)}{n}f^n(a)\right)\right) \cdots \right)\right)$$

Operations with Truncated Forms of Infinite Series

Advanced analysis often involves the numerical evaluation of truncated series. Generally, the approach is to truncate the series at something on the order of four terms and use the series to evaluate the function over the region that has a good fit with the function being considered. Once a series is generated, whether with Chebyshev polynomials, Taylor series, the binomial series, Legendre polynomials, or some other means, such operations can be performed on the series as inverting the series, taking the square root of it, squaring it, multiplying or dividing it, taking the exponential of it, or taking the logarithm of it. This is conveniently done by manipulating the coefficients in the series. These operations are tabulated in Table 3-1 for the three series

$$s_1 = 1 + a_1 x + a_2 x^2 + a_3 x^3 + a_4 x^4 + \cdots$$

$$s_2 = 1 + b_1 x + b_2 x^2 + b_3 x^3 + b_4 x^4 + \cdots$$

$$s_3 = 1 + c_1 x + c_2 x^2 + c_3 x^3 + c_4 x^4 + \cdots$$

Among convenient series manipulations is the reversion of series, where the dependent variable is solved in terms of the independent variable. Given the series

$$y = ax + bx^2 + cx^3 + dx^4 + ex^5 + fx^6 + \cdots$$

we can write x as a function of y as

$$x \approx Ay + By^2 + Cy^3 + Dy^4 + Ey^5 + Fy^6 + \cdots$$

where

$$A = \frac{1}{a}$$

$$B = -\frac{b}{a^3}$$

$$C = \frac{2b^2 - ac}{a^5}$$

$$D = \frac{5abc - a^2 d - 5b^3}{a^7}$$

$$E = \frac{6a^2 bd + 3a^2 c^2 + 14b^4 - 21ab^2 c - a^3 e}{a^9}$$

$$F = \frac{7a^3 be + 7a^3 cd + 84ab^3 c - a^4 f - 28a^2 bc^2 - 42b^5 - 28a^2 b^2 d}{a^{11}}$$

Table 3-1 Series Operations

Operation	c_1	c_2	c_3	c_4
$s_3 = s_1^n$	na_1	$\frac{1}{2}(n-1)c_1a_1 + na_2$	$c_1a_2(n-1) + \frac{c_1a_1^2}{6}(n-1)(n-2) + na_3$	$na_4 + c_1a_3(n-1) + \frac{1}{2}n(n-1)a_2^2 + \frac{1}{2}(n-1)(n-2)c_1a_1a_2 + \frac{1}{24}(n-1)(n-2)(n-3)c_1a_1^3$
$s_3 = s_1 s_2$	$a_1 + b_1$	$b_2 + a_1b_1 + a_2$	$b_3 + a_1b_2 + a_2b_1 + a_3$	$b_4 + a_1b_3 + a_2b_2 + a_3b_1 + a_4$
$s_3 = s_1/s_2$	$a_1 - b_1$	$a_2 - (b_1c_1 + b_2)$	$a_3 - (b_1c_2 + b_2c_1 + b_3)$	$a_4 - (b_1c_3 + b_2c_2 + b_3c_1 + b_4)$
$s_3 = e^{(s_1-1)}$	a_1	$a_2 + \frac{a_1^2}{2}$	$a_3 + a_1a_2 + \frac{a_1^3}{6}$	$a_4 + a_1a_3 + \frac{a_2^2}{2} + \frac{a_2a_1^2}{2} + \frac{a_1^4}{24}$
$s_3 = 1 + \ln(s_1)$	a_1	$a_2 - \frac{a_1c_1}{2}$	$a_3 - \frac{(a_2c_1 + 2a_1c_2)}{3}$	$a_4 - \frac{(a_3c_1 + 2a_2c_2 + 3a_1c_3)}{4}$

Transformation of Series

Occasionally, slow-converging series are encountered in numerical analysis where the object is to compute the sum of the series to high accuracy. Usually we would use some form of economization to improve the accuracy of such a series. We may, however, also know another series that can be used to improve the convergence (accuracy) of the original series. This is convenient when numerically evaluating the sum of a slowly converging series of the form

$$s = \sum_{k=0}^{\infty} a_k$$

where it is known that the series does in fact converge and where we have another series

$$c = \sum_{k=0}^{\infty} c_k$$

which is also convergent and which we know to have the sum c and the limit of a_k / c_k as k approaches infinity to equal λ (where λ is not equal to zero); then

$$s = \lambda c + \sum_{k=0}^{\infty} \left(1 - \lambda \frac{c_k}{a_k} \right) a_k$$

This technique is known as Kummer's transformation. It transforms one series into another that is more convenient for numerical evaluation. While not developed originally for this purpose, it turns out to be quite useful in numerical evaluation of slowly converging series.

Another approach to numerically evaluating a truncated series is to use the Euler-Maclaurin summation formula. This is another technique for numerically evaluating series using another series that converges more quickly. Provided that the difference of derivatives at the end points of the interval over which the series is being evaluated is small, the Euler-Maclaurin summation formula is

$$s = \sum_{k=1}^{n-1} f_k \cong \int_0^n f(k)\,dk - \tfrac{1}{2}(f_0 - f_n) + \tfrac{1}{12}\left(f_n^{(I)} - f_0^{(I)} \right)$$

$$- \tfrac{1}{720}\left(f_n^{(III)} - f_0^{(III)} \right) + \frac{\left(f_n^{(V)} - f_0^{(V)} \right)}{30240}$$

3-5 THE SOLUTION OF POLYNOMIALS

The numerical solution of a polynomial on the 5100 requires a clear understanding of the possible location of the polynomial's roots in the complex plane. For this reason, we take a few moments to refresh our understanding of algebraic equations. It should be remembered that an nth-order algebraic equation has n roots. If the coefficients in the polynomial are real, the roots of the equation are either all real, some being equal and some not, or have pairs of roots that are complex conjugates of each other and other roots that are real with various locations on the real axis. The occurrence of complex roots in complex conjugate pairs arises from our assumption that the coefficients in the polynomials are real, not complex. If the coefficients are complex, of course, the roots can occur anywhere in the complex plane. In this book we concern ourselves only with polynomials that have real coefficients, since they are the most frequently encountered algebraic equations in engineering analysis.

The Solution of Quadratic Equations

If we are given a quadratic equation of the form

$$az^2 + bz + c = 0$$

its roots can be numerically evaluated with the formula

$$z_1 = -\left(\frac{b}{2a}\right) + \frac{\sqrt{q}}{2a}$$

$$z_2 = -\left(\frac{b}{2a}\right) - \frac{\sqrt{q}}{2a}$$

where

$$q = b^2 - 4ac$$

From time to time we will make use of the following easily verified properties of the roots:

$$z_1 + z_2 = -\frac{b}{a}$$

$$z_1 z_2 = \frac{c}{a}$$

It is apparent from the equations for the two roots that

1. If $q > 0$, the two roots will be real and unequal.
2. If $q = 0$, the two roots are both real and equal.

3. If $q < 0$, the roots occur in complex conjugate pairs.

The numerical evaluation on the calculator should involve first the calculation to determine q and then the use of equations for the roots for their evaluation once the situation of the roots is determined.

Solution of Cubic Equations

If we are given a cubic equation of the form

$$z^3 + a_2 z^2 + a_1 z + a_0 = 0$$

the first step in computing its roots is to calculate q and r:

$$q = \frac{a_1}{3} - \frac{a_2^2}{9}$$

$$r = \frac{a_1 a_2 - 3a_0}{6} - \frac{a_2^3}{27}$$

Then:

1. If $q^3 + r^2 > 0$, the cubic equation has one real root and a pair of complex conjugate roots.
2. If $q^3 + r^2 = 0$, all the roots are real and at least two are equal.
3. If $q^3 + r^2 < 0$, all roots are real and unequal (the irreducible case).

Once the nature of the roots is known, it is a simple matter to use the following equations to evaluate the roots on the calculator. First, compute

$$s_1 = \left[r + (q^3 + r^2)^{1/2} \right]^{1/2}$$

$$s_2 = \left[r - (q^3 + r^2)^{1/2} \right]^{1/2}$$

Then the roots can be calculated from an understanding of their nature and the following three equations:

$$z_1 = (s_1 + s_2) - \frac{a_2}{3}$$

$$z_2 = \frac{-(s_1 + s_2)}{2} - \frac{a_2}{3} + \frac{i\sqrt{3}}{2}(s_1 - s_2)$$

$$z_3 = -\frac{(s_1 + s_2)}{2} - \frac{a_2}{3} - \frac{i\sqrt{3}}{2}(s_1 - s_2)$$

Note that if $q^3 + r^2 = 0$, s_1 will equal s_2 and the imaginary component of the roots will drop out, leaving the two z roots, z_2 and z_3, equal, while z_1 may not necessarily be equal, depending on the value of s_2.

Once the roots of the cubic equation are evaluated, they satisfy the following relations:

$$z_1 + z_2 + z_3 = -a_2$$

$$z_1 z_2 + z_1 z_3 + z_2 z_3 = a_1$$

$$z_1 z_2 z_3 = -a_0$$

These relations can be used as a check on the calculation of the roots.

The process of numerically evaluating the roots of the quartic equation is somewhat involved, even for pocket calculator evaluation. Under some conditions, however, simple evaluations can be made. For example, consider the quartic equation

$$z^4 + a_3 z^3 + a_2 z^2 + a_1 z + a_0 = 0$$

One approach to evaluating the roots of this quartic equation is to find the real root of the cubic equation

$$\mu^3 - a_2 \mu^2 + (a_1 a_3 - 4a_0)\mu - (a_1^2 + a_0 a_3^2 - 4a_0 a_2) = 0$$

and then determine the four roots of the quartic equation as solutions to the two quadratic equations

$$v^2 + \left[\frac{a_3}{2} - \left(\frac{a_3^2}{4} + \mu_1 - a_2\right)^{1/2}\right]v + \frac{\mu_1}{2} - \left[\left(\frac{\mu_1}{2}\right)^2 - a_0\right]^{1/2} = 0$$

$$v^2 + \left[\frac{a_3}{2} + \left(\frac{a_3^2}{4} + \mu_1 - a_2\right)^{1/2}\right]v + \frac{\mu_1}{2} + \left[\left(\frac{\mu_1}{2}\right)^2 - a_0\right]^{1/2} = 0$$

Once the roots of the quartic are evaluated and can be written in the form

$$z^4 + a_3 z^3 + a_2 z^2 + a_1 z + a_0 = (z^2 + p_1 z + q_1)(z^2 + p_2 z + q_2)$$

the following conditions hold:

$$p_1 + p_2 = a_3$$

$$p_1 p_2 + q_1 + q_2 = a_2$$

$$p_1 q_2 + p_2 q_1 = a_1$$

$$q_1 q_2 = a_0$$

Finally, if z_1, z_2, z_3, z_4 are the roots of the quartic equation, the following conditions hold among the roots:

$$z_1 + z_2 + z_3 + z_4 = -a_3$$

$$\sum z_i z_j z_k = -a_1$$

$$\sum z_i z_j = a_2$$

$$z_1 z_2 z_3 z_4 = a_0$$

Again, these conditions can be used to check on the calculation of the roots.

The evaluation of the roots of a polynomial up to quartics is tedious and usually inaccurate (at best) on a slide rule, by hand analysis, or even on the old mechanical calculators (though accurate); it is a relatively fast and accurate process on the pocket calculator, however.

3-6 SUCCESSIVE APPROXIMATION METHODS

Here, we are concerned wth the problem of determining the roots of an equation, but the equation is of a more general form. We are looking for the condition

$$f(x) = 0$$

That is, we are looking for the values of x such that $f(x)$ will equal zero. In this case $f(x)$ need not be a polynomial in x. If we let $x = x_n$, the approximation of the root, then when f_n is not equal to 0 it is equal to ϵ (the error). If we now use ϵ to update our estimate of the root,

$$\Delta x = c_n \epsilon_n = c_n f_n$$

we can write

$$x_{n+1} = x_n + c_n f(x_n), \qquad (n = 1, 2, 3 \cdots) \tag{3-1}$$

When it is found that $f'(x)$ is greater than or equal to zero and the constants c_n are negative and bounded, the sequence of x_n converges monotonically to the root $x = r$. If c is a constant less than zero and f' is greater than zero, the process converges but not necessarily monotonically. A number of approaches have been developed to compute c_n. Among these are the *regula falsi* method, the method of successive iterations, Newton's method, and the Newton-Raphson method. The *regula falsi* method begins with the assumption that we are given $y = f(x)$; the objective is to find $x = r$ such that $f(r) = 0$. We choose a pair of values of x, x_0, and x_1 such that $f(x_0)$ and $f(x_1)$ have opposite signs. Then equation 3-1 can take the form

$$x_2 = x_1 - \left(\frac{x_1 - x_0}{f_1 - f_0} \right) f_1 = \frac{f_1 x_0 - f_0 x_1}{f_1 - f_0} \tag{3-2}$$

The third- and higher-order estimates of the root x_n are computed using x_2 and either x_0 or x_1 for which $f(x_0)$ or $f(x_1)$ is of opposite sign to $f(x_2)$.

This method is equivalent to an inverse interpolation. This is apparent from the form of equation 3-2.

In the method of successive iterations, the approach is to write the equation in an implicit form and use successive iterations to solve the equation $x = F(x)$. The iteration scheme is to compute

$$x_{n+1} = f(x_n)$$

The sequence of solutions to this implicit equation will converge to a zero of $x = F(x)$ if there exists a q such that

$$|f'(x)| \leqslant q < 1 \qquad \text{for} \quad a \leqslant x \leqslant b$$

and

$$a \leqslant x_0 \pm \frac{|f(x_0) - x_0|}{1 - q} \leqslant b$$

This is an attractive method for use on the 5100 because it does not involve remembering special formulas such as those associated with the *regula falsi* or the Newton (Newton-Raphson) methods. The problem encountered in applying the method of successive iterations on the implicit form of the equation whose roots are to be determined is that the implicit equation may not converge as quickly as other methods based on additional information (such as the derivatives of $f(x)$) whose function it is to ensure rapid convergence of the method.

Newton's method is to compute recursively estimates of the roots of the function $f(x)$ using the formula

$$x_{n+1} = x_n - \frac{f(x_n)}{f'(x_n)} \qquad (3\text{-}3)$$

where $x = x_n$ is an approximation to the solution, $x + r$, of $f(x) = 0$. The sequence of solutions generated with Newton's rule will converge quadratically to $x = r$. The condition for monotonic convergence is that the product $f(x_0)f''(x_0)$ is greater than zero, and $f'(x)$ and $f''(x)$ do not change sign in the interval (x_0, r). The conditions for oscillatory convergence are also straightforward. When the product $x(x_0)f''(x_0)$ is less than zero and $f'(x)$ and $f''(x)$ do not change sign in the interval (x_0, x_1), equation 3-3 will converge, though it will oscillate. These conditions only hold, of course, when

$$x_0 \leqslant r \leqslant x_1$$

When Newton's method is applied to the evaluation of nth roots, we find that given $x^n = N$, if x_k is an approximation of $x = N^{1/n}$ then a sequence of improved x_k can be generated:

$$x_{k+1} = \frac{1}{n}\left(x_k\left(\frac{N}{x_k^n} + n - 1\right)\right)$$

This method will converge quadratically to x for all n. It is derived here to show the procedure:

1. We wish to compute $x = (N)^{1/n}$.
2. Form $f(x) = (x^n - N) = 0$ from (1).
3. For Newton's rule, $x_{k+1} = x_k - f(x_k)/f'(x_k)$, we need $f(x_k)$ and $f'(x_k)$.
4. $f(x_k) = (x_k^n - N)$ and $f'(x_k) = (nx_{k}^{n-1} - 0)$.
5. Substituting the results of (4) into (3) we find (6).
6.

$$x_{k+1} = x_k - \left(\frac{x_k^n - N}{nx_k^{n-1}}\right) = \frac{nx_k}{n} - \frac{x_k^n + N}{nx_k^{n-1}}$$

$$= \frac{1}{n}\left[\frac{N}{x_k^{n-1}} + (n-1)x_k\right] = \frac{1}{n}\left(x_k\left(\frac{N}{x_k^n} + n - 1\right)\right)$$

3-7   REFERENCE

For further details on finding zeros of polynomials, analytic substitution and advanced formulas such as Bessel functions, Legendre polynomials, and Chebyshev polynomials, see *Scientific Analysis on the Pocket Calculator*, J. M. Smith, 2nd ed., Wiley, New York (1977).

CHAPTER 4

NUMERICAL INTEGRATION

4-1 INTRODUCTION

This chapter is about numerical integration on the 5100. The concepts of numerical integration are covered in detail with emphasis on those numerical integrators that are sized for use on the 5100. Do not overlook the examples in this chapter, because they have been developed to illustrate certain key numerical integration concepts.

There are basically two types of integral with which we are concerned here: the definite integral and the indefinite integral. The definite integral is given by the formula

$$y(b) = y(a) + \int_a^b f(x)\,dx$$

and the indefinite integral is defined by

$$y(x) = y(a) + \int_a^x f(t)\,dt$$

The definite integral is characterized by computing the area under the curve of a bounded function; the indefinite integral can be thought of as integrating differential equations or *solving* differential equations.

4-2 DEFINITE INTEGRATION

Computing the area under an arbitrary curve is usually based on the concept of analytic substitution. The idea is to use a known function whose definite integral is easily evaluated to substitute for the arbitrary function to be integrated. The integration is actually performed on the substitute function and attributed to the integral of the arbitrary function to the degree to which it approximates the latter. In classical mathematics the substitute functions to be integrated are usually polynomials. The polynomial is then analytically integrated and, insofar as the polynomial approximates the continuous function, the integral is attributed to the integral of the arbitrary function. When the integrand is a polynomial of degree n and the approximating function is also a polynomial of degree n, the formula can be made exact by appropriately selecting the coefficients in the integration formula.

The process of *analytic substitution* or of other means of approximating definite and indefinite integrals is so fascinating that virtually every numerical analyst finds new ways to rederive many of the classical formulations and a few others as well. The focus here remains on classical developments, which are straightforward and easy to apply on the 5100 scientific calculator. The reader should be aware, however, of the tremendous quantity of good mathematics in numerical integration developed in the past 20 years. These new developments are a result of the introduction of the digital computer only a few decades ago and because new developments were needed to cope with problems in new technological areas. Structures, communication systems, control systems, design of aircraft, and the design of chemical plants are areas where the simulation of systems with widely separated eigenvalues and the numerical integration of functions that are almost neutrally stable (at large integration step size) have produced new integration concepts based on the technology to which they were being applied. Structural dynamicists have developed special numerical integration formulas for integrating their "stiff differential equations." Controls analysts have produced such formulas based solely on frequency-domain considerations. And special single-step real-time numerical integration formulas have been developed by simulation scientists.

The most easily understood numerical integrators that are easily programmed on the 5100 are the Euler, trapezoidal, and midpoint integrators.

4-3 EULER INTEGRATION

If we approximate the $f(x)$ on a bounded interval $a \leqslant x \leqslant b$ by a constant, we can write the equation for the approximating function over the interval as

$$y(x) = f(a)$$

Then

$$\int_y^b (x)\,dx \simeq f(a)(b-a)$$

This simple relation was first studied by the great mathematician Euler, from whom it gets it name.

Example 4-1 Numerically integrate the function $\sin\theta$ over the interval [0, 360°].

Approach:

Step 1 Evaluate $\sin\theta$ every 10°.

Step 2 Use Euler numerical integration to integrate $\sin\theta$ over the 10° interval.

Step 3 Program the 5100 to iterate steps 1 and 2.

Analysis:

Step 1 Let $\theta_N = \theta_{N-1} + \Delta\theta$; $\theta_{-1} = -10°$.

Step 2 $X_N = X_{N-1} + 10°\left(\dfrac{2\pi\,\text{rad}}{360°}\right)\sin\theta_N$; $X_{-1} = 0$.

Step 3 Let X be stored in A.

Let θ be stored in B.

Let $\Delta\theta$ be stored in C.

Program your 5100 in the AER mode as follows:

1; B = B + C STO B,   A = A + C × (2π ÷ 360) × sin B STO A

Then initialize the program in the *COMP* mode as follows:

$$0 \text{ STO A}$$
$$-10 \text{ STO B}$$
$$+10 \text{ STO C}$$

Set the trig functions to read arguments in degrees.

Now we can fill in the table of integration as follows:

Ans 1	Ans 2
θ	X
0	0
10°	0.030
20°	0.090
30°	0.177
40°	0.289
50°	0.423
60°	0.574
70°	0.738
80°	0.910
90°	1.085
180°	1.995
270°	0.910
360°	0.000

As a check case, the values at 0°, 90°, 180°, 270°, and 360° should be

$$X = \int_0^\theta \sin\theta \, d\theta = 1 - \cos\theta$$

which

$$\text{at } \theta = 0°; \quad I = 0$$
$$\text{at } \theta = 90°; \quad I = 1$$
$$\text{at } \theta = 180°; \quad I = 2$$
$$\text{at } \theta = 270°; \quad I = 1$$
$$\text{at } \theta = 360°; \quad I = 0$$

4-4 TRAPEZOIDAL INTEGRATION

If we approximate the function $f(x)$ on a bounded interval $a \leqslant x \leqslant b$ by a line through the end points, we can write the equation for the approximating function over the interval as

$$y(x) = f(a) + \left[\frac{f(b) - f(a)}{b - a} \right](x - a)$$

$$y(x) = \frac{(b - x)f(a) + (x - a)f(b)}{b - a}$$

Integrating, we find

$$\int_a^b y(x)\,dx = \left(\frac{f(b)+f(a)}{2}\right)(b-a)$$

This equation computes the area under the straight-line interpolation between two end points. This is called trapezoidal integration because the area is enclosed by a trapezoid formed by lines connecting the end points, the abscissa, and the vertical lines connecting the end points to the abscissa. If the interval is large, the trapezoidal approximation can lead to large numerical integration error. This is resolved by a repeated application of the trapezoidal rule on smaller intervals of the dependent variable. When this is done for equally spaced intervals, Δx, trapezoidal integration takes the form

$$\int_a^b f(x)\,dx = \Delta x \left(\frac{f(a)}{2} + f(a+\Delta x) + f(a+2\Delta x) + \cdots + \frac{f(b)}{2}\right)$$

Euler and trapezoidal integration, though not the most accurate, are straightforward to apply and are easily remembered.

4-5 ERROR IN NUMERICAL INTEGRATION

We do not here aim to explore the derivation of integration or error formulas—merely to tabulate the commonly used ones and put them in a form that is immediately useful for analysis with the 5100. Nevertheless, it is instructive to examine the error of a simple integration formula as a means for understanding the error equations given for the more sophisticated integration formulas. Following Hamming, then, we examine the truncation error in the trapezoidal integration algorithm by substituting a Taylor series expansion into the integration formula. By comparing both sides of the results, we can then determine the error associated with the analytic substitution process in the numerical integration. Specifically, if we write the integrand in its Taylor series expanded form as

$$f(x) = f(a) + (x-a)f'(a) + \frac{(x-a)^2}{2!}f''(a) + \cdots$$

and substitute this into both sides of the trapezoidal integration formula, we find that, on integration, the left side becomes

$$\frac{(b-a)}{1!}f(a) + \frac{(b-a)^2}{2!}f'(a) + \frac{(b-a)^3}{3!}f''(a) + \cdots$$

The right side becomes

$$\frac{\Delta x}{2}\left[f(a)+(b-a)f'(a)+\frac{(b-a)^2}{2}f''(a)+\cdots+f(a)\right]+\epsilon$$

where $\Delta x=(b-a)$. After canceling like terms on both sides we can derive the truncation error formula:

$$\epsilon+\frac{(b-a)^3}{4}f''(a)+\cdots=\frac{(b-a)^3}{3!}f''(a) \tag{4-1}$$

$$\epsilon=\left(\frac{1}{3!}-\frac{1}{4}\right)(b-a)^3f''(a)-\frac{(b-a)^4}{5}f'''(a)-\cdots$$

If we assume the largest part of the error term to be given by the first term in its series expansion, we can write

$$\epsilon\approx-\frac{(b-a)^3f''(a)}{12}$$

or, more generally,

$$\epsilon\approx-\frac{(b-a)^3f''(\theta)}{12},\qquad(a\leqslant\theta\leqslant b) \tag{4-2}$$

Similarly, for Euler integration,

$$\epsilon\simeq\frac{(b-a)^2}{2}f'(\theta),\qquad(a\leqslant\theta\leqslant b)$$

If, however, the function has contributions to the error formula that are large in the higher-order terms, these error formulas do not apply. It is applicable for many of the practical engineering and scientific problems, and thus is generally quoted as the error associated with trapezoidal integration.

Figure 4-1 shows that for "concave-up" functions the error in trapezoidal integration is always slightly on the high side; for "concave-down" functions it is slightly on the low side. You should select intervals so that, *to the eye*, the errors accumulated over one set of intervals may cancel the errors over another set of intervals. We can extend the error formula for simple trapezoidal integration to the composite formula by

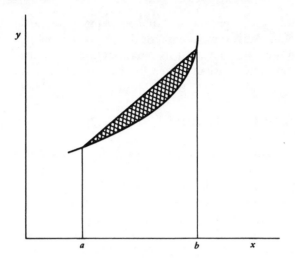

Figure 4-1 Truncation error in trapezoidal integration.

similar reasoning. What results is

$$\epsilon \approx -\frac{(b-a)\Delta x^2}{12}f''(0), \qquad (a \leqslant 0 \leqslant b) \qquad (4\text{-}3)$$

To evaluate these error formulas requires evaluation of the derivatives of f at some value θ. The question is, what θ? One of many approaches is to find the derivatives of the function being considered, compute the minimum error and the maximum error and divide by 2 to obtain the average value of the derivative of interest over the interval. The question is, what criterion for numerically evaluating the error formula should be used for a particular problem? Unfortunately, there is no easy answer. From an engineering viewpoint, the error defined by equation 4-1 perhaps has more meaning than those such as 4-2 and 4-3 as are most often quoted in numerical analysis books. The process of deriving the error formula may be more useful than giving some generalized error formula, in that the engineer or scientist can derive the error formula that is suited to his specific problem.

Another aspect of the numerical error formulas associated with integration formulas is that they are absolute errors, whereas the error of interest is usually relative error. Again, the author has no easy solution of the problem of deriving relative error formulas for numerical analysis. The difficulty is pointed out here to warn the student or first-time numerical analyst about error formulas in general. Preferably he should derive his own formula for a particular problem being numerically analyzed. An estimate of the error in a numerical approximation over an analytical calculation must be made, but its interpretation is not straightforward and the result cannot be casually given from questionable error formulas.

4-6 MIDPOINT INTEGRATION

Midpoint integration uses the midvalue of an interval and the derivative of the integrand evaluated at the midvalue to define the slope at the midpoint of the interval, again forming a trapezoid whose area under the curve approximates that of a function to be considered.

The midpoint integration formula as developed by Hamming is easy to follow and nicely introduces the concept of a general approach to deriving polynomial approximations for analytic substitution. We are to derive an integration formula of the form

$$\int_a^b f(x)\,dx = w_1 f\left(\frac{a+b}{2}\right) + w_2 f'\left(\frac{a+b}{2}\right)$$

Hamming's weighting coefficients can be easily derived by noting that we first require this formula to be exact for $f(x) = 1$. This gives

$$b - a = w_1$$

We also require that this formula be exact for $f(x) = x$, which leads to

$$\frac{b^2 - a^2}{2} = w_1\left(\frac{a+b}{2}\right) + w_2$$

We can determine the two Hamming's coefficients by solving these equations simultaneously:

$$w_2 = \left(\frac{b^2 - a^2}{2}\right) - \frac{(b-a)(a+b)}{2} = \left(\frac{b^2 - a^2}{2}\right) - \left(\frac{b^2 - a^2}{2}\right) = 0$$

$$w_1 = b - a$$

We therefore find the midvalue integration formula to be

$$\int_a^b f(x)\,dx = (b-a)f\left(\frac{a+b}{2}\right) \tag{4-4}$$

We see that midpoint integration developed in this manner results in rectangular integration. That is, the area formed by the rectangle sampled at the midvalue is identically equal to the area under the tangent line at the midvalue of the interval. What is even more surprising is that errors introduced by midpoint integration are less than errors introduced by trapezoidal integration. At first it might seem paradoxical that the low-order rectangular integration could be as good as trapezoidal integration (that formulas based on a single point of f could be as accurate as a two-point trapezoidal formula). In fact, rectangular integration can be made as precise as desired if the sample point on a bounded interval can be varied until the mean value theorem of calculus is satisfied. *Once again*, rectangular integration can be made as precise as the true integral provided that the point at which the function is sampled on the interval can be determined, so that the rectangle formed by the sampled value and the lines connecting the end points of the function on the interval and the abscissa itself have the same area as that under the function bounded on the interval. The attentive reader will recognize that this is simply the mean-value theorem of integral calculus stated in words. The T-integrator, described later, is *designed* to numerically implement the mean-value theorem.

Again, to find the truncation error term, we use the Taylor series:

$$f(x) = f(a) + \frac{(x-a)}{1!}f'(a) + \frac{(x-a)^2}{2!}f''(a) + \cdots$$

where, upon substituting in both sides of equation 4-4, we find

$$\epsilon + \frac{(b-a)^3}{8}f''(a) + \cdots = \frac{(b-a)^3}{6}f''(a) + \cdots$$

This is usually simplified to

$$\epsilon \approx \frac{(b-a)^3 f''(a)}{24} \tag{4-5}$$

or, more generally, for $a \leqslant \theta \leqslant b$

$$\epsilon \approx \frac{(b-a)^3 f''(\theta)}{24}$$

Comparing equations 4-5 and 4-2, we see that midpoint rectangular integration is more accurate than endpoint trapezoidal integration even though the rectangular integration is based on knowing the function at only one point while the trapezoidal rule of integration requires the knowledge of the function at two end points.

Extending the midpoint integration formula, we find, as in the trapezoidal formula, the *composite* midpoint integration formula to be of the form

$$\int_a^b f(x)\,dx$$

$$= \Delta x \left[f\left(a + \frac{\Delta x}{2}\right) + f\left(a + \frac{3\Delta x}{2}\right) + f\left(a + \frac{5\Delta x}{2}\right) + \cdots + f\left(b - \frac{\Delta x}{2}\right) \right] + \epsilon$$

where its error formula is given by

$$\epsilon \approx \frac{(b-a)\Delta x^2}{24} f''(\theta), \qquad (a \leqslant \theta \leqslant b)$$

Note also that extended trapezoidal integration can be modified to include end points outside the interval $[a, b]$. The modified trapezoidal rule is given by

$$\int_a^b f(x)\,dx \cong \Delta x \left[\frac{f(a)}{2} + f(a + \Delta x) + f(a + 2\Delta x) + \cdots + \frac{f(b)}{2} \right]$$

$$+ \frac{\Delta x}{24} \left[-f(a - \Delta x) + f(a + \Delta x) + f(b - \Delta x) - f(b + \Delta x) \right]$$

where the error associated with modified trapezoidal integration is given by

$$\epsilon = \frac{11(b-a)\Delta x^4}{720} f''''(\theta), \qquad (a + \Delta x) \leqslant \theta \leqslant (b + \Delta x)$$

which is usually much more accurate than extended midpoint integration with only slightly more work.

4-7 OTHER POPULAR DEFINITE INTEGRATION FORMULAS

Simpson's rule, perhaps the most commonly used integration formula, is given by

$$\int_0^{2\Delta x} f(x)\,dx \cong \frac{\Delta x}{3}(f_0 + 4f_1 + f_2)$$

Its associated error formula is given by

$$\epsilon = -\frac{\Delta x^5}{90} f''''(\theta), \qquad (0 \leq \theta \leq 2\Delta x)$$

Simpson's rule has the nice property that it integrates cubics exactly even though it samples only three points of the integrand and in addition has very small error terms when Δx is less than 1 and on the order of one-half.

Simpson's rule can also be extended (on an even number of intervals) according to the formula

$$\int_{x_0}^{x_{2n}} f(x)\,dx = \frac{\Delta x}{3}(f_0 + 4f_1 + 2f_2 + 4f_3 + 2f_4 + \cdots + f_{2n})$$

Its error formula is given by

$$\epsilon = \frac{n\Delta x^5}{90} f''''(\theta), \qquad (x_0 \leq \theta \leq x_0 + 2n\Delta x)$$

Similarly, the three-eights rule for definite integration is given by

$$\int_{x_0}^{x_3} f(x)\,dx = \frac{3\Delta x}{8}(f_0 + 3f_1 + 3f_2 + f_3)$$

Its associated error formula is

$$\epsilon = -\frac{3\Delta x^5}{80} f''''(\theta), \qquad x_0 \leq \theta \leq x_3$$

Finally, the simplest extended integration formula is the Euler-Maclaurin formula:

$$\int_{x_0}^{x_n} f(x)\,dx = \Delta x\left(\frac{f_0}{2} + f_1 + f_2 + f_3 + \cdots + \frac{f_n}{2}\right) - \left(\frac{B_2 \Delta x^2}{2!}\right)(f_n' - f_0') - \cdots$$

$$- \left(\frac{B_{2k}\Delta x^{2k}}{2k!}\right)(f_n^{(2k-1)} - f_0^{(2k-1)}) + \epsilon_{2k}$$

It has the error formula

$$\epsilon_{2k} = \left\{ \frac{\theta n B_{2k+2}\Delta x^{(2k+3)}}{(2k+2)!}\right\}\left\{ \max_{x_0 \leq x \leq x_n} |f(x)^{2k+2}|\right\}, \qquad (-1 \leq \theta \leq 1)$$

Here B_{2k} is a Bernoulli number.

4-8 INDEFINITE NUMERICAL INTEGRATION

Indefinite numerical integration is the numerical method for solving differential equations. Given the equation

$$\frac{dy}{dx} = f(x,y)$$

we would usually solve it by indefinite integration as follows:

$$y = y_0 + \int_{x_0}^{x} f(t,y)\,dt$$

It is apparent that the solution of the differential equation depends on its own evaluation of the integral. This is precisely the chief problem in indefinite integration; that is, indefinite integrals are in an implicit form.

Note that an explicit indefinite integral takes the form

$$y = y_0 + \int_{x_0}^{x} f(t)\,dt$$

which is a special case of the differential equation

$$\frac{dy}{dx} = f(x)$$

Clearly, this type of numerical integration can be performed analytically, hence is not of concern here.

The simplest indefinite numerical integration algorithm is Euler's integration formula:

$$y_{n+1} = y_n + \Delta x \left(\frac{dy}{dx} \right)_n$$

Here we see that a new estimate (y_{n+1}) of y is based on the old estimate (y_n) and its derivative $[(dy/dx)_n]$. The derivative is usually calculated directly from the differential equation once y is estimated. Since the new estimate y_{n+1} is based on the old estimate y_n' and the old value y_n it is clearly an "open-loop" process where the new value y_n is based on an extrapolation from previously known data and thus is subject to extrapolation errors. The process of determining new values of y is really a simple extension of determining the *direction field* associated with a solution of a differential equation. In general, the approach is to start at some initial condition (x_0, y_0) and calculate the slope, using the differential equation:

$$y_0' = f(f_0, y_0)$$

One then moves an interval Δx in the direction of the slope to a second point, which we now regard as the new initial point, and repeat the process iteratively. If small enough steps are taken we can reasonably hope that the sequence of solution values given by this procedure will lie close to the solution of the differential equation. In general, all of the elements of solving differential equations using numerical indefinite integration are present here. A table of the values of x, y, y', and Δy must be computed at each step in the numerical integration process. Also, the problem must be defined by specifying not only the differential equation and its initial conditions, but also the interval over which it is desired to solve the equation. It is then possible to select a convenient integration interval, and an integration formula that is accurate for that interval. For example,

$$\frac{dy}{dx} = e^{-y} - x^2$$

with initial conditions $y = 0$, $x = 0$. When integrated with Euler's integration formula

$$y_n = y_{n-1} + \Delta x y_{n-1}'$$

requires a specification of the interval Δx. The simplest approach is to experimentally determine the Δx that will accurately (as judged by the analyst) integrate the differential equation. Consider solutions of this differential equation with $\Delta x = 0.05$, 0.1, 0.2, and 0.3. Try this yourself. Program your 5100 to solve this nonlinear differential equation using an Euler integrator.*

*Try this if you had any problem:

where

$$x \text{ is stored in A}$$
$$dy/dx \text{ is stored in B}$$
$$y \text{ is stored in C}$$
$$\Delta x \text{ is stored in D}$$

Table 4-1 Solution of $dy/dx = e^{-y} - x^2$

	Exact Solution	Euler Integrated Solution			
x	y	$\Delta x = 0.05$	$\Delta x = 0.10$	$\Delta x = 0.2$	$\Delta x = 0.3$
0.0	0.0				
0.1	0.09498	0.09694	0.09900	—	—
0.2	0.17977	0.18261	0.18557	0.19200	—
0.3	0.25389	0.25672	0.25964	—	0.27300
0.4	0.31667	0.31872	0.32077	0.32506	—
0.5	0.36731	0.36786	0.36833	—	—
0.6	0.40488	0.40329	0.40152	0.39756	0.39333
0.7	0.42839	0.42407	0.41942	—	—
0.8	0.43686	0.42923	0.42119	0.40395	—
0.9	0.42929	0.41782	0.40582	—	0.35277
1.0	0.40477	0.38895	0.37264	0.33749	—

The results are set out in Table 4-1. A comparison of the numerically integrated solutions with the exact solution shows that the sensitivity of the solution's accuracy depends strongly on the integration step size. This is true, in general, for all numerical integrators when the integration step size is even a reasonable fraction of the "response time" of the differential equation.*

A disadvantage of the Euler method is that it introduces systematic phase shift or lag (extrapolation) errors at each step. The procedure can be modified (modified Euler integration) to give better results—that is, greater accuracy for essentially the same method and the same amount of work.

4-9   THE MODIFIED EULER INDEFINITE INTEGRATION METHOD

An alternative to introducing lag into the calculation is to arrange the sampling so that the integrand is sampled not at the end point of the interval over which the integration is taking place but at the midpoint. This is similar to the development of the midpoint trapezoidal formula. The task is to perform the integral

$$\int_{x_{n-1}}^{x_{n+1}} y'(x)\,dx$$

---

*Approximately the time required to move from one equilibrium condition to another.

using the midpoint formula. We wish to predict the next value of y based on present and past values of the independent variable. The midvalue prediction leads to

$$p_{n+1} = \bar{y}_{n-1} + 2\Delta x y'_n$$

Using this predicted value, we can now compute the slope at the predicted solution point, by way of the differential equation,

$$p'_{n+1} = f(x_{n+1}, p_{n+1})$$

and then apply the trapezoidal rule developed previously to update the estimate of the predicted solution point:

$$y_{n+1} = y_n + \frac{\Delta x}{2}(p'_{n+1} + y'_n)$$

The correction is called the corrected value of y_{n+1}. It is apparent that we are using the average of the slopes at the two end points of the interval of integration as the average slope in the interval.

In summary, this method has three steps:

Step 1 Predict the value of y_{n+1}, given the formula

$$p_{n+1} = y_{n-1} + 2\Delta x y'_n$$

Step 2 Compute the derivative at the predicted value, using the differential equation that describes the system:

$$p'_{n+1} = f(x_{n+1}, p_{n+1})$$

Step 3 Make a second estimate of the value of y_{n+1}, using trapezoidal integration:

$$y_{n+1} = y_n + \frac{\Delta x}{2}(y'_n + y'_{n+1})$$

This process of prediction and correction has led to the naming of this type of integration as the predict-correct concept of numerical integration.

4-10 STARTING VALUES

In our previous analysis we assumed that we had values for the dependent
and independent variables at the starting or initial point. However, the
algorithm requires not only starting values, but also earlier values. The
previous values can be obtained in two ways. They can be computed on the
5100 or they can be analytically hand calculated. Both methods will be
presented here.

The hand calculation method is based on the use of the Taylor series
expansion of the function:

$$y(x+\Delta x)=y(x)+\Delta x y'(x)+\frac{\Delta x^2}{2}y''(x)+\cdots$$

The derivatives to be evaluated in the Taylor series expansion can be
found from the differential equation by repeated differentiation. The
number of terms of course depends on the step size and the accuracy
desired. But, again, these are matters that can all be easily evaluated on the
pocket calculator and the number of terms required can be empirically
determined by continuing to take them until the desired accuracy is
achieved.

The method for machine calculation is based on repeated use of the
corrector formula. Again, if we are given the initial point (x_0, y_0), we can
estimate the earlier point (x_{-1}, y_{-1}) by way of the "unmodified" Euler
integration, working backwards as follows:

$$x_{-1}=x_0-\Delta x$$

$$y_{-1}=y_0-\Delta x y_0' \qquad \text{(first estimate of } y_{-1})$$

We can use the estimate of the previous value of y combined with the
differential equation to evaluate the derivative at the previous value of y.
The trapezoidal corrector formula can then be repeated to iteratively
correct the previous estimate until it achieves the accuracy desired for the
calculation. The system of equations for the correction process become

$$y'_{-1}=f(x_{-1},y_{-1}) \qquad\qquad \text{(first estimate of } y'_{-1})$$

$$y_{-1}=y_0-\frac{\Delta x}{2}(y_0'+y'_{-1}) \qquad \text{(second estimate of } y_{-1})$$

$$y'_{-1}=f(x-1,y_{-1}) \qquad\qquad \text{(second estimate of } y'_{-1})$$

$$\vdots \qquad\qquad\qquad\qquad\qquad\qquad \vdots$$

If, after a few iterations, the previous value of y does not stabilize, the integration step size can be halved, the previous value of $y_{n-1/2}$ computed, and the process repeated to compute y_{n-1}. Another alternative is to use the value of $y_{n-1/2}$ to estimate the value of $y_{n+1/2}$, the process is repeated to take a half step forward to y_{n+1}, and then these values used as the starting values for the predict-correct integration algorithm.

4-11   T-INTEGRATION

T-Integration (tunable integration) is a new flexible integration concept that permits the integration formula to be tuned to the system of equations it is solving. In its simplest form it is written:

$$y_n = y_{n-1} + \lambda T\left[\gamma \dot{y}_n + (1 - \gamma)\dot{y}_{n-1}\right]$$

This equation is based on adjusting the phasing of the integration so as to satisfy the mean value theorem (as opposed to numerical integration algorithms based on analytical substitution techniques). The parameter γ controls the amount of transport lead (or lag) imposed on the integrand of the integral. For example, $\gamma = -\frac{1}{2}$ means that the integrand has been time delayed one full sample period, while $\gamma = +\frac{3}{2}$ implies that the integrand is time advanced one sample period. Since in the numerical integration of differential equations the solution point is not known before it is computed, and thus cannot be made part of the integral, it is estimated using an extrapolation formula. It is apparent from the equation for this integrator that the weight of the two coefficients in the integrand is performing the interpolation/extrapolation operation. Therefore, an approximate equivalent form of the T-integration equation is

$$y_n = y_{n-1} + \lambda T\left[(\gamma + 1)\dot{y}_{n-1} - \gamma \dot{y}_{n-2}\right]$$

In its application, T-integration is usually used in the following manner:

1. The differential equation and the interval over which its solution is to be computed are defined together with its initial condition.
2. The integration step size is set equal to one-tenth of the interval size or one-tenth of the shortest period in the oscillations of the solution expected for the differential equation, whichever is smaller. If the solution of the differential equation is expected to be exponential, or smooth, or of monotonic nature, the step size is set at one-tenth the interval over which the solution is to be evaluated.

3. If the integration is an open-loop process, that is, the integrand is not a function of the integral, then γ is set equal to $\frac{1}{2}$ and the differential equation is numerically integrated. If, however, the integrand is a function of the integral, γ is set equal to $\frac{3}{2}$. When closed-loop integration is performed, the sequence of solutions can be plotted and the points connected with straight lines.

4. The solution at the starting γ can then be compared with check cases that may be run at smaller integration step sizes or with empirically prepared check cases. γ is varied and the integral repeated iteratively until a value of γ is found that results in the least error. Then the T-integrator is said to be tuned to the problem it is solving. Surprisingly, once γ is found for a given set of differential equations, it is found to be relatively invariant with different initial conditions and with different forcing functions; even when the differential equations are moderately nonlinear.

Note, however, that the *dynamics* of the solution prepared with the T-integrator match the *dynamics* of any check case. That is, although the T-integrator, which is a low-order integrator, permits accurate simulation of the *dynamics* of a discrete process, it does so at the sacrifice of a slight phase error (time shift). Nevertheless, in many engineering applications it is sufficient for determining, for example, peak overshoot, natural frequency, damping, resonant frequencies, and conditions of dynamic instability, which are the purpose of the analysis. In general, it must be remembered that all of the integration formulas presented here are usually not for generating of numbers to six places, but rather for solving problems and understanding the dynamics of processes for the purposes of design, test, and evaluation, or all of them.

Numerical Integration of a Second-Order Differential Equation

Example 4-2   Solve the equation

$$\ddot{X} + 2(0.7)(2\pi)\dot{X} + (2\pi)^2 X = f(t)^*$$

This DEQ describes a 1-Hz, 0.7 damped oscillator.

Approach:

Step 1   Solve for the highest derivative in the DEQ.

Step 2   Numerically integrate each derivative.

---

*For this problem $\zeta = 0.7$, $\omega_n = (2\pi \text{ rad})/\text{sec} = 1$ Hz. The coefficient for the $\dot{X}$ term is $2\zeta\omega_n$. The coefficient for the X term is ω_n^2.

Step 3   Program the 5100 to iterate steps 2 and 3.
Here

$$\left.\begin{array}{l} X_0 = 0 \\ \dot{X}_0 = 0 \end{array}\right\} \text{initial conditions}$$

$$f(t) = 4\pi^2 U(t); \text{ the unit step function } = U(t)$$

Analysis:

Step 1   Solve for $\ddot{X}$.

$$\ddot{X} = 4\pi^2 - 2.8\pi\dot{X} - 4\pi^2 X$$

Step 2   Numerically integrate, using Euler's integration algorithm, $\ddot{X}$ and $\dot{X}$.

$$\dot{X}_N \simeq T\ddot{X}_N + \dot{X}_{N-1}$$

$$X_N \simeq T\dot{X}_N + X_{N-1}$$

Now, let $T = 0.05$ (not too accurate, but stable).

Step 3   Program the 5100 to keep track of X, $\dot{X}$, $\ddot{X}$, T and N.
Note:

> let X be stored in A
> let $\dot{X}$ be stored in B
> let $\ddot{X}$ be stored in C
> let T be stored in D
> let N be stored in E   (start at $N=0$)

AER mode

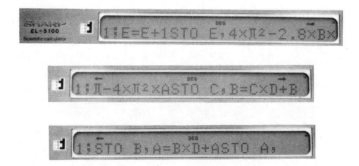

COMP mode

0.0 STO A STO B STO C STO E

0.05 STO D

The first stroke of "comp" yields N

The second stroke of "comp" yields $\ddot{X}$

The third stroke of "comp" yields $\simeq \dot{X}$

The fourth stroke of "comp" yields $\simeq X$

Now we can fill in the table.

Ans 1	Ans 2	Ans 3	Ans 4	
N	$\ddot{X}$	$\dot{X}$	X	
0	39.48	0	0	From the initial conditions
2	18.22	2.88	0.24	
4	−3.61	2.93	0.54	
6	−9.39	2.07	0.78	$\omega_n = 1$ Hz
8	−8.40	1.18	0.91	$\zeta = 0.7$
10	−5.58	0.55	0.98	
12	−3.04	0.19	1.01	
14	−1.33	0.02	1.02	
16	−0.40	−0.04	1.01	
18	0.01	−0.05	1.01	
20	0.14	−0.03	1.00	

Now you try your hand at finding the response of a second-order system with $\omega = 1$ Hz and $\zeta = 0.3$. The results for Euler integration are tabulated as follows:

Ans 1	Ans 2	Ans 3	Ans 4
N	$\ddot{X}$	$\dot{X}$	X
0	39.48	0	0
2	28.14	3.38	0.27
4	4.85	4.43	0.70
6	−12.18	3.58	1.09
8	−19.15	1.78	1.31
10	−17.09	−0.03	1.35
12	−9.78	−1.21	1.26
14	−1.57	−1.57	1.10
16	4.39	−1.26	0.97
18	6.78	−0.61	0.89
20	6.00	0.02	0.88
22	3.40	0.43	0.91

| Ans 1 | Ans 2 | Ans 3 | Ans 4 |
N	$\ddot{X}$	$\dot{X}$	X
24	0.50	0.55	0.96
26	− 1.58	0.44	1.01
28	− 2.40	0.21	1.04
30	− 2.11	− 0.01	1.04
32	− 1.18	− 0.15	1.03
34	− 0.16	− 0.20	1.01
36	− 0.57	− 0.15	1.00
38	0.85	− 0.07	0.99
40	0.74	0.01	0.98

4-12   REFERENCES

See J. M. Smith, "Recent Developments in Numerical Integration," *ASME Journal of Dynamic Systems, Measurement, and Control*, March 1974, pages 61–70, and *Scientific Analysis on the Pocket Calculator*, 2nd ed., Wiley (1977).

CHAPTER 5

LINEAR SYSTEMS SIMULATION

5-1 INTRODUCTION

The analysis of linear constant coefficient systems is important because they are frequently encountered in the design of continuous processes. The dynamic characteristics of a linear system's response to known types of forcing functions are usually studied when setting the parameters for a system design. In this chapter we discuss the synthesis of recursion formulas by which the response of a linear dynamic process to sampled values of its forcing function can be conveniently computed. We tailor numerical integration and other discrete approximation methods for computing the dynamics of continuous processes to pocket calculator analysis. On the Sharp 5100 it is much easier to iterate a recursion formula to compute the dynamics of a process than it is to actually conduct the numerical integration of the process. Under some circumstances (when there are no hard nonlinearities, such as limits, hysteresis, and dead zones), it is quite easy to develop the recursion formulas from the integration formulas, thus eliminating many steps in the computing of the solution to high-order differential equations. In fact, the number and size of the algebraic formulas used can be reduced as much as 80% with recursion formulas (difference equations) as compared to those needed in direct numerical integration of a differential equation.

5-2 DERIVATION OF DIFFERENCE EQUATIONS BY NUMERICAL INTEGRATION SUBSTITUTION

Examples of many numerical integration formulas have already been discussed, such as Euler's integration formula, mid-value Euler integration, trapezoidal integration, and T-integration. We used the differential equation to numerically evaluate the derivatives at the initial condition and then from the starting values in the integration formulas we predicted the solution to the differential equation in the neighborhood of the initial conditions.

Another use of a numerical integration formula is to form a difference equation. Consider the first-order constant coefficient differential equation

$$\tau \dot{x} + x = Q$$

where

$$x = x(t)$$

$$Q = Q(t)$$

$$\tau = \text{a constant}$$

Now consider the Euler integration formula

$$x_n = x_{n-1} + T\dot{x}_{n-1}$$

We can solve for the rate in the integration formula, using the differential equation, as follows:

$$\dot{x}_{n-1} = \frac{1}{\tau}(Q_{n-1} - x_{n-1})$$

This can be substituted back into the numerical integration formula

$$x_n = x_{n-1} + \frac{T}{\tau}(Q_{n-1} - x_{n-1})$$

which, when simplified, gives the difference equation

$$x_n = \left(1 - \frac{T}{\tau}\right)x_{n-1} + \frac{T}{\tau}Q_{n-1}$$

This recursion formula computes, for example, the 100th step in the solution of the differential equation on the basis of data generation on the 99th step. The indices in the recursion formula keep track of the iteration that is being computed when solving the differential equation. They also indicate the approximate time at which the solution value will compare with $x(t)$, that is, $t = nT$ if the solution begins at $T \cong 0$. We shall see later that $t \neq nT$, but it is sufficiently close to approximately label the time in the sequence of solution values of the difference equation.

The use of recursion formulas in solving differential equations has two advantages:

1. They reduce the size and number of algebraic expressions needed to evaluate the solution of the differential equation on the 5100.
2. In linear constant coefficient processes they permit the use of *implicit* integration formulas.

It is these formulas in which the rates of a state variable are a function of the state itself. The trapezoidal integration formula is an example:

$$x_{n+1} = x_n + \frac{T}{2}(\dot{x}_{n+1} + \dot{x}_n)$$

Trapezoidal integration computes the $n+1$ value of x based on the $n+1$ value of $\dot{x}$. However, evaluating $\dot{x}$ in the differential equation requires x_{n+1}. This results in an implicit equation, whose solution is a function of itself. When implicit integration formulas are used to derive difference equations, the implicit equation can be solved *algebraically*. For example, consider the implicit Euler integration (rectangular integration), which takes the form

$$x_n = x_{n-1} + T\dot{x}_n$$

By way of our first-order differential equation, we obtain

$$\dot{x}_n = \frac{1}{\tau}(Q_n - x_n)$$

which, when substituted back into the implicit rectangular integration formula, gives the difference equation

$$x_n = x_{n-1} + \frac{T}{\tau}(Q_n - x_n)$$

Note that this equation is still in implicit form; that is, x_n is a function of itself. However, it can be solved algebraically as follows:

$$x_n + \frac{T}{\tau}x_n = x_{n-1} + \frac{T}{\tau}Q_n$$

$$\left(1 + \frac{T}{\tau}\right)x_n = x_{n-1} + \frac{T}{\tau}Q_n$$

$$\therefore x_n = \left(\frac{1}{1 + T/\tau}\right)x_{n-1} + \left(\frac{T/\tau}{1 + T/\tau}\right)Q_n$$

Let us now compare the implicit and explicit Euler difference equations from the standpoints of numerical stability, numerical error, the manner in which the differential equation seeks its final value, and their implementation on the calculator.

The stability of these first-order difference equations is completely determined by the magnitude of the first coefficient in the difference equation. That is, if the term

$$\frac{1}{1+T/\tau} \qquad\qquad 1+\frac{T}{\tau}$$

Implicit integration   Explicit integration

exceeds ± 1, the difference equation becomes unstable. For example, if $a=2$ in the difference equation $y_n = ay_{n-1}$, the difference equation takes on the solution values shown in Table 5-1. Note, however, that at $a=0.9$ the difference equation is stable, as shown in Table 5-2. The stability criterion in first-order difference equations generally is that the magnitude of a be less than or equal to 1. Now, notice the first-order difference equation that is generated with the explicit Euler integration.

Our aim here is to determine the conditions under which the integration step size and the system's time constant allows a stable difference equation, rather than leading to numerical instability. We, therefore, first determine the conditions under which the magnitude of a is less than or equal to 1. That is

$$\left|1-\frac{T}{\tau}\right| \leqslant 1$$

Solving the inequality for T/τ, we see that the region of stability for the

Table 5-1 Unstable Response of the Difference Equation $y_n = ay_{n-1}$ where $a=2$

n	y_n
1	1
2	2
3	4
4	8
5	16
.	.
.	.
.	.

Table 5-2 Stable Response of the Difference Equation $y_n = aY_{n-1}$ **where** $a = 0.9$

n	y_n
1	1
2	0.9
3	0.81
4	0.729
5	0.6561
.	.
.	.
.	.

difference equation derived with explicit Euler integration is

$$0 \leqslant \frac{T}{\tau} \leqslant 2$$

On examining the difference equation derived with rectangular integration (implicit Euler integration), we see that the condition under which

$$\left| \frac{1}{1 + T/\tau} \right| \leqslant 1$$

is

$$0 \leqslant \frac{T}{\tau}$$

Clearly, the difference equation developed with rectangular integration is much more stable than that generated by Euler explicit integration. This is a specific example of the more general result that implicit integration of constant coefficient linear differential equations leads to intrinsically more stable difference equations than do those developed with explicit integration formulas. We therefore concentrate on the use of implicit integration formulas in developing difference equations for simulating continuous processes.

Now, let us look at the accuracy of these simulating difference equations. Consider the following example.

Example 5-1 Compare explicit and implicit Euler numerical integration of a differential equation.

Approach:

Step 1 Substitute each of these numerical integrators into the differential equation so as to derive a difference equation that describes numerically integrated motion.

Step 2 Program the difference equations on the 5100 and compare the results of solutions to the difference equation with the exact solution to the differential equation.

Analysis:

For this example we will numerically integrate the linear, first-order differential equation

$$\tau \dot{x} + x = Q(t)$$

where

$$\tau = \text{the response time in seconds}$$
$$Q = U(t); \text{ the unit step function}$$

Step 1 From the text we know the difference equation that describes the explicitly integrated differential equation is

$$x_n = \left(1 - \frac{T}{\tau}\right) x_{n-1} + \frac{T}{\tau} Q_{n-1}$$

and the implicitly integrated differential equation is

$$x_n = \left(\frac{1}{1 + T/\tau}\right) x_{n-1} + \left(\frac{T/\tau}{1 + T/\tau}\right) Q_n$$

In both formulas $Q_n = Q_{n-1} = 1$. The exact solution of this well-known differential equation is

$$x = (1 - e^{-nT/\tau})$$

Step 2 We can program these algebraic formulas in the AER mode as

1; A = A + 1 STO A,
    B = (1 − C) × B + C STO B,
    D = (D + C) ÷ (1 + C) STO D,

    G = 1 − e − (A × C) STO G,

$$H = G - B \text{ STO } H,$$
$$I = G - D \text{ STO } I,$$
$$E = H \div G,$$
$$F = I \div G,$$

Here

A = (n) the number of iterations through the difference equation, initially zero

B = x_n explicit; B initially set to $x_0 = 0$

C = T/τ; precomputed and stored in C and initially zero

D = x_n implicit; D initially set to $x_0 = 0$

H = the absolute error in explicit numerical integration

I = the absolute error in implicit numerical integration

E = the percentage error in decimals in explicit numerical integration

F = the percentage error in decimals in implicit numerical integration

G = x_n exact.

In the *COMP* mode set

$$0 = A = B = D = E = F = G = H = I$$
$$1.5 = C = \frac{T}{\tau}$$

Now, by repeated *COMP* key strokes, we can fill in Table 5-3. This data is summarized in Table 5-4 and shows the sequence of solution values for the explicit and implicit difference equation's response to a unit step. The greatest precision is clearly achieved with the implicit formula. These

Table 5-3 Comparison of Implicit and Explicit Integration-Derived Difference Equations when $T/\tau = 1.5$

① Ans 1	② Ans 2	③ Ans 3	④ Ans 4	⑤ Ans 5	⑥ Ans 6	⑦ Ans 7	⑧ Ans 8
n Step No.	x_n Explicit	x_n Implicit	x_n Exact	Absolute $4-2$	Errors $4-3$	Percentage $5 \div 4$	Errors $6 \div 4$
1	1.500	0.600	0.777	-0.723	0.177	-9.31	2.28
2	0.750	0.840	0.950	$+0.200$	0.110	$+2.11$	1.16
3	1.125	0.936	0.989	-0.136	0.053	-1.38	0.53
4	0.938	0.974	0.998	$+0.060$	0.023	$+0.60$	0.23

Table 5-4 Comparison of Implicit and Explicit Integration-Derived Difference Equations when $T/\tau = 1.5$

Normalized Time	Exact $x(nT)$	Implicit $x(nT)$	Implicit %Error	Explicit $x(nT)$	Explicit %Error
$\dfrac{T}{\tau} = 0$	0	0	0	0	0
$\dfrac{T}{\tau} = 1.5$	0.776	0.600	+2.28	1.50	−9.31
$\dfrac{T}{\tau} = 3.0$	0.950	0.840	+1.16	0.75	+2.11
$\dfrac{T}{\tau} = 4.5$	0.990	0.936	+0.53	1.125	−1.38
$\dfrac{T}{\tau} = 6.0$	0.998	0.974	+0.23	0.938	+0.60

difference equations were tested for an integration step size divided by the time constant equal to $\frac{3}{2}$, which challenges the stability of the Euler-derived difference equation. Both equations appear to be stable. However, the implicit difference equation is obviously more accurate than is the explicit equation. This is another special case of a general property of difference equations derived with implicit integration to simulate linear constant coefficient systems. The implicitly derived difference equations are generally more accurate than those derived explicitly.

Finally, let us examine the steady state that all these difference equations achieve. To do so, we must examine the nonhomogeneous equation (since in a homogeneous equation all the end conditions of the steady states approach zero, thus making comparison impossible). For the continuous and discrete equations, the step response has the forms shown below:

$$y = Q(1 - e^{-t/\tau}), \quad y_n = \left(1 - \frac{T}{\tau}\right)y_{n-1} + \frac{T}{\tau}Q_{n-1}$$
$$\text{Exact} \qquad\qquad\qquad\qquad \text{Explicit}$$

$$y_n = \left(\frac{1}{1 + T/\tau}\right)y_{n-1} + \left(\frac{T/\tau}{1 + T/\tau}\right)Q_n$$
$$\text{Implicit}$$

In the steady state

$$x_n = x_{n-1}$$

Thus we can write the final value as follows:

$$\lim_{t\to\infty} y(t)=Q \qquad y_n=Q_n=Q_{n-1} \qquad y_n=Q_n$$
$$\text{Exact} \qquad\qquad \text{Explicit} \qquad\qquad \text{Implicit}$$

In summary: both the explicitly and implicitly derived difference equations achieve the same final value for the unit step forcing functions which is the final value for the true continuous process. But an implicitly derived recursion formula is more stable and more accurate than its explicitly derived counterpart.

These recursion formulas are particularly useful in evaluating a system's response to an arbitrary forcing function. Provided that the integration step size is small compared with the largest period of interest in the oscillations of the forcing function, the recursion formulas can efficiently evaluate the system's response to an arbitrary forcing function on the 5100 where the implicit difference equations take up much less memory than do the numerical integration formulas and direct numerical integration.

A possible difficulty associated with the implicit integration formula for evaluating the response of a system to an arbitrary forcing function is its assumption that the forcing function is known at time nT, in order to compute the response of the system at time n. If the forcing function is of the form

$$f=f(y,t)$$

the evaluation of f_n requires

$$f_n=f(y_n,t_n)$$

but since the difference equation is still to compute y_n, it is not yet in our table of solution values; instead we have only a tabulated value for y_{n-1}. In this case we can use an extrapolation formula to estimate y_n by way of the two past values, or we can use y_{n-1} merely as an approximation of y_n. This can be done when the forcing function's components are (from a Fourier analysis viewpoint) of lower frequency than is the natural frequency of the system described by the differential equation. To achieve this, we calculate a few values of the difference equation, assuming in evaluating the forcing function that $y_n\approx y_{n-1}$ and generating the first few terms of the forcing function f, and use a difference table to evaluate whether f is changing rapidly. If the change is rapid, we simply use an interpolation formula to make a first estimate of y_n based on y_{n-1} and y_{n-2}. The author rarely finds it necessary to use the extrapolation scheme in the practical evaluation of the solution to differential equations.

This technique of deriving difference equations to simulate continuous dynamic processes is extremely useful for simulating the dynamics of nonlinear processes. One problem is that most implicit difference equations cannot be solved for nonlinear differential equations. That is, the implicit equation is a nonlinear equation, and usually only iterative techniques can be used to solve it. However, the explicit difference equation is easily derived and easily put in a form that can be quickly evaluated on the pocket calculator, as opposed to numerically integrating the nonlinear equation.

5-3 STABLE DIFFERENCE EQUATIONS

Recursion formulas for simulating continuous dynamic processes can also be derived by assuming a difference equation of the same order as the differential equation to be simulated. Then match the roots of the difference equation with the roots of the differential equation and include an "adjustment factor" so as to match the final value of the difference equation with the final value of the differential equation. All that remains, then, is to add another "adjustment factor" to match the phasing of the difference equation to the phasing of the solution to the differential equation. For example, consider again the simple first-order constant coefficient continuous process

$$\tau \dot{x} + x = Q$$

Assume that this equation has a solution of the homogeneous equation

$$x = e^{ts}$$

On substitution, we can derive the indicial equation as

$$(\tau s + 1)e^{ts} = 0$$

which has the characteristic root

$$s = -\frac{1}{\tau}$$

Clearly, then, the solution to the homogeneous differential equation takes the form

$$x = c_1 e^{-t/\tau}$$

The solution to the nonhomogeneous equation can be derived with the convolution integral where the solution of the homogeneous equation is convolved with the forcing function:

$$x = \int_0^t Q(k) e^{[(k-t)/\tau]} dk$$

The complete solution to the differential equation then takes the form

$$x = e^{-t/\tau} \left\{ \int_0^t Q(k) e^{k/\tau} dk + c_1 \right\}$$

Similar procedures can be followed for higher-order differential equations, using either time-domain analysis, Laplace transform theory, or even Z-transform theory.

Let us assume that we are going to simulate this continuous process with a difference equation whose roots and final value match those of the continuous process. We assume a difference equation:

$$x_n = a x_{n-1} + b Q_n$$

A solution to the homogeneous difference equation is of the form

$$x_n = c_1 e^{-nT/\tau}$$

Upon substitution, it leads to the indicial equation for the difference equation:

$$c_1 e^{-nT/\tau} (1 - a e^{T/\tau}) = 0$$

Thus for the roots of the difference equation to match the roots of the differential equation, we require that

$$a = e^{-T/\tau}$$

This determines the coefficient in the difference equation that accomplishes the pole matching between the difference and differential equations. It is clear that the solution to the homogeneous difference equation is

$$x_n = e^{-T/\tau} x_{n-1}$$

This procedure has now guaranteed that the dynamics of the difference equation will match the dynamics of the differential equation because their roots are equivalent and they will generate equivalent solution values as seen by the exponential decay of both. What remains is to compute the final value of the difference equation and match it with that of the differential equation. The procedure here is more straightforward in that the nonhomogeneous difference equation takes the form

$$x_n = e^{-T/\tau} x_{n-1} + b Q_{n-1}$$

where the steady state of the root-matched difference equation is achieved when

$$Q_n = Q_{n-1}$$

$$x_n = x_{n-1}$$

Then

$$x_n = \frac{b}{1 - e^{-T/\tau}} Q_{n-1}$$

By including the final value adjustment factor

$$b = 1 - e^{-T/\tau}$$

we can make the difference equation achieve the same final value as the differential equation. Thus the simulating difference equation takes the form

$$x_n = e^{-T/\tau} x_{n-1} + (1 - e^{-T/\tau}) Q_{n-1} \qquad (5\text{-}1)$$

Notice that the homogeneous solution of this difference equation matches the homogeneous solution of the differential equation *exactly*. Also, it generates a sequence of solutions that are exact for the step response $(Q(t) = U(t))$ and will generate solutions that are a good approximation of the differential equation's response to an arbitrary forcing function. Also notice that this difference equation is incapable of going unstable, regardless of the integration step size (because the term $e^{-T/\tau}$ is always less than 1 no matter how big T gets provided that $\tau > 0$).

From the tabulated values it may appear that there is significant error in the solutions generated with the dynamics-matched difference equation and that generated with the continuous differential equation. However, equation 5-1 makes it clear that the difference equation solutions are lagging the continuous solutions. The dynamics are usually identical to the differential equation except for this effect of phase shift. Of course, we could reduce the step size to bring the two curves closer, but this is not an efficient or correct approach to reducing this kind of error. Or we can compensate for this phasing error (transport delay) by determining with interpolation at what time the sequence of solutions generated by the difference equation matches the differential equation and then including that transport time in the tabulation of the sequence of solutions generated in the difference equation. Suppose that we know that for the fourth entry in a table of solution values the true continuous solution lies somewhere between $t = 3T$ and $4T$. Using inverse interpolation, we can determine the time at which the discrete solution matches the continuous solution and then arbitrarily select that time as the reference time from which we count nT intervals.

It is important to remember that the solution values generated with difference equations and even with numerical integration formulas are operating at a problem time which is different from the sequence of times nT. That is, problem time in a discrete approximation of a continuous time process is different from the sequence of values nT. Hence the indices in the recursion formulas represent the number of iterations, not time nT. The analyst must determine the actual timing of the sequence of solution values in order to compare them with a true continuous-time check case. It is the author's experience that many engineers and programmers, on large digital computers as well as on pocket calculators, overlook this problem of timing and try to compare continuous and discrete computing processes at times nT instead of recognizing that numerical integration is *an approximating process*. There is a timing problem also in the synthesis of simulating difference equations by dynamics matching. In fact, discrete systems are different in their operation on the flow of information in feedback loops, whether in numerical integrators or in difference equations. Thus the phasing of the two sequences of values between continuous and discrete dynamic processes must be taken into account by the analyst. The problem really arises only with large integration step sizes, but it is precisely then that efficiency is at a premium and, especially on the pocket calculator, workload is substantially reduced from that for an integration step size only half as long.

Once again we find the pocket calculator may be the analytical tool for teaching the difference between discrete and continuous systems dynamics and the simulation of one with the other.

A few of the commonly encountered linear processes and their simulating difference equations using dynamics matching methods are tabulated in Table 5-5.

It is *imperative* that when the simulating difference equations are used the table of solution values be referenced to the number of iterations through the difference equations, not to time nT. The comparison of discrete solution values with a continuous check case *involves timing considerations*, and it is the analyst's responsibility to determine the proper comparison in a manner similar to that mentioned above.

Table 5-5 Difference Equations for Commonly Encountered Linear Constant Coefficient Systems

$G(s)$	Difference Equation for Simulation
$\dfrac{y}{x} = \dfrac{1}{\tau s + 1}$	$y_n = e^{-T/\tau} y_{n-1} + (1 - e^{-T/\tau}) x_n$
$\dfrac{y}{x} = \dfrac{\tau s}{\tau s + 1}$	$\begin{cases} z_n = e^{-T/\tau} z_{n-1} + (1 - e^{-T/\tau}) x_n \\ y_n = x_n - z_n \end{cases}$
$\dfrac{y}{x} = \dfrac{w_n^2}{s^2 + 2\zeta w_n s + w_n^2}$	$\begin{cases} B = e^{-2\zeta w_n T} \\ A = 2e^{-\zeta w_n T} \cos\left\{ w_n T (1 - \zeta^2)^{1/2} \right\}; \quad 0 < \zeta < 1 \\ y_n = A y_{n-1} - B y_{n-2} + (1 - A + B) x_n \end{cases}$
$\dfrac{y}{x} = \dfrac{s(s + 2\zeta w_n)}{s^2 + 2\zeta w_n s + w_n^2}$	$\begin{cases} z_n = A z_{n-1} - B z_{n-2} + (1 - A + B) x_n; \quad 0 < \zeta < 1 \\ y_n = x_n - z_n \\ \text{For } A \text{ and } B \text{ see above} \end{cases}$

The difference equations just developed by dynamics matching methods have some very important general properties. These difference equations are intrinsically stable if the process under consideration is stable. That is, there is no sample period T to cause these equations to become unstable if the continuous process that they are simulating is stable. This is because the roots of the differential equation are matched with the roots of the difference equation; hence if the continuous process is stable, the discrete process is stable independent of sample period. Showing that the magnitude of the roots of the discrete system are less than or equal to 1 will prove this; the very way in which they are formulated shows this to be so. For example, in the first-order case that we just developed, when the roots of the discrete system are matched to the roots of the continuous system, the discrete system root is given by

$$e^{-T/\tau}$$

which will be always less than or equal to 1 for all $T > 0$ and for $\tau > 0$. The only condition on using the difference equation is that the forcing function be sampled at a rate equal to twice the highest frequencies of interest in the forcing function.

Also, the final value of the discrete difference equation will always match that of the continuous difference equation, independent of sampling period and without the final value adjustment factor. That this is so can be established by the fact that in the steady state the present and past values of the response of the system are the same. When substituted into the difference equation, the final value of the response can be computed in terms of the input forcing function, which is found to match the final value of the continuous dynamic process being simulated.

There is a limitation in the use of these simulating difference equations. Clearly, a second-order continuous system can have three different dynamic characteristics: when the two roots of the system are real and equal; when they are real and unequal; and when both are complex. The dynamics of the second-order continuous system with complex roots is damped oscillatory in nature, and the response of the system with real roots is nonoscillatory, being damped only. Each case requires different types of difference equations. It is important, then, to know where the roots are in the complex plane to determine which difference equation is to be used. This is particularly true if the coefficients in the differential equation are changing with time and are not fixed, as in the case of linear constant coefficient systems. When the coefficients are time varying, these difference equations can be used for piecewise linear constant coefficient approximation, with the results matching closely the numerically integrated solution of the time-varying differential equation. However, if the time-varying roots jump on and off the real axis, switching from one difference equation to another is necessary. That is, one difference equation simulates the dynamics of the process when the roots are real but not equal; another difference equation simulates the dynamics when the roots are real and equal; and yet another difference equation serves when the roots are complex. The choice of the appropriate set of difference equations is fairly straightforward, but note that the implicit difference equation generated in Section 5-2 does not require this changing of difference equations and thus might be more applicable from the standpoint of quickly simulating continuous processes on the calculator.

5-4 REFERENCE

For additional material on Fourier analysis and optimization, see *Scientific Analysis on the Pocket Calculator*, J. M. Smith, 2nd ed., Wiley, New York (1977).

CHAPTER 6

STATISTICS AND PROBABILITY

6-1 INTRODUCTION

It is fairly easy to quantify and define an individual. You can measure his or her weight, height, waist circumference, and so on. Additionally, you can give a name. But what about a *group* of individuals? How do you quantify and define a *group* of individuals in some useful way? For that matter, what are the mathematical properties of *groups*? Answers to these kinds of intriguing questions are the domain of STATISTICS. Many of the answers are well known. They are based on certain "commonsense" definitions. Those presented here are working definitions and are *not* given in abstract mathematical notation.

A *data population* is simply a collection of unorganized data called the raw data. Arrays and frequency distribution are ways of organizing the data, so that the statistics of a collection of data can be determined.

Arrays are arrangements of raw data in ascending or descending order; that is, the data are tabulated starting with the largest number and proceeding to the smallest, or vice versa. The *range* of an array is the difference between the largest and smallest numbers in the array.

When large sets of data are stratified into categories and the number of elements in the data set belong to each category or class, we form a *data distribution*. This is done by generating a table of data by category or class, together with the class *frequencies*, that is, the number of elements of the set of all data belonging to each class. Such a tabular array is called a frequency distribution or frequency table. An example appears in Table 6-1. Here the classes or categories are the intervals of height. Data arranged in a frequency distribution are often also called grouped data. The term "class mark" refers to the midpoint of a class interval.

In general, frequency distributions are developed by first determining the range of the raw data, dividing the range into a convenient number of

Table 6-1   Heights of 100 Male Students in a
University

Height (in.)	Number of Students	Cumulative Number of Students
60–63	4	4
63–66	18	22
66–69	41	63
69–72	28	91
72–75	9	100

class samples of the same size, and then determining the number of observations that fall into each class interval (this, by definition, is called the class frequency). Once the frequency distribution is known, *histograms* can be developed to visualize the frequency distribution. Histograms are simply a plot of the frequency against the range of raw data. An example of a histogram for the frequency distribution in Table 6-1 is shown in Figure 6-1.

The relative frequency of a class is its frequency divided by the total frequency of all classes. It is multiplied by 100 to obtain a percentage. Plots of relative frequency over the range of the data are called a percentage distribution or relative frequency distribution.

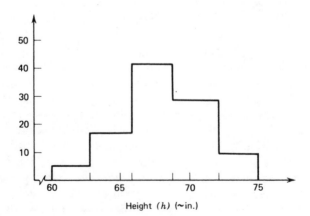

Height (h) (~in.)

Figure 6-1   Frequency distribution (number of students in height interval Δh).

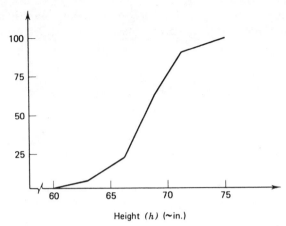

Height (h) $(\sim$in.$)$

Figure 6-2 Cumulative frequency distribution (cumulative number of students with heights less than or equal to h).

The cumulative frequency distribution is simply defined as the total frequency of all data less the upper class or category boundary of a given class interval. The third column of Table 6-1 shows the cumulative frequency distribution of the height of the 100 students. A cumulative frequency distribution can be plotted over the range of data as shown in Figure 6-2.

Relative cumulative frequency distributions are defined like the frequency distributions, and so is their percentage.

Were the data to increase without bound, the histogram's frequency distribution and cumulative frequency distribution would be expected to be developed with increasingly finer quantitization until smooth curves are obtained. The analysis of probability using continuous frequency distribution functions is the field of probability analysis. Here we concentrate on the statistics associated with finite sized data sets. We examine probability as well, but our emphasis is on the statistical analysis of small sized data sets that can be reasonably analyzed with the 5100 calculator.

6-2 MEASURES OF CENTRAL TENDENCY

The mean, median, mode, and other *measures of central tendency* are the *statistics* of data distributions. Numbers that describe the centroid of the distributions are called measures of central tendency. There are many such measures, all called averages. The most common is the arithmetic mean.

Two other measures of central tendency that are important in statistics are the median and the mode. The arithmetic mean is defined by the relationship

$$\overline{X} = \frac{\sum\limits_{j=1}^{N} X_j}{N} = \frac{\sum X}{N}$$

The arithmetic mean is quickly evaluated on the Sharp 5100 in the statistical mode by inputting the raw data with the *data* key. 2nd F $\overline{x}$ calculates the mean.

If certain numbers occur more than once, in particular with frequencies $f_1, f_2, \ldots, f_n$, then the arithmetic mean is defined by

$$\overline{X} = \sum\limits_{j=1}^{K} f_j X_j / \sum\limits_{j=1}^{K} f_j = \frac{\sum fX}{\sum f} = \frac{\sum fX}{N}$$

Remember that on the 5100, the frequencies are input as multipliers on the number and the proper order is $X \times f$. The frequency is the second term in the product. If the data are input in the reverse order, information of N, the *size* of the raw data set, will be in error.

The *median* of an array of numbers (i.e., numbers arranged in order of increasing or decreasing size) is defined to be the middle value of the array. If the array has an odd number of elements, there is a single middle

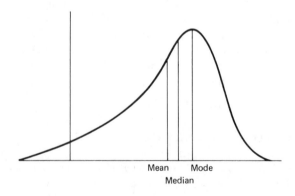

Mean   Mode

Median

Figure 6-3   Mean, median, and mode of a typical symmetric distribution.

value. If the array has an even number of elements, there are two middle values, in which case the median is defined as the average of the two middle values.

The *mode* of a distribution is that value which occurs most frequently and is the peak of the histogram.

In general, distributions can have more than one mode, but only one mean and one median.

For unimodal distributions that are only slightly asymmetrical (skewed) the mean, median, and mode are approximately related as

$$\text{mean} - \text{mode} = 3(\text{mean} - \text{median})$$

The distinction between the mean, median, and mode is shown in Figure 6-3.

6-3 MEASURES OF DISPERSION

Dispersion is defined to be the distribution or spread of the data about the average value. It is also frequently called the variation of the data. The number of measures of dispersion or variation of data about the mean is almost as large as the number of great statisticians. Here we are concerned only with the *standard deviation* of the data, which is denoted by s and is defined by the relations

$$s = \left[\frac{\sum_{j=1}^{N} \left(X_j - \overline{X} \right)^2}{N-1} \right]^{1/2} = \left(\frac{\sum \left(X - \overline{X} \right)^2}{N-1} \right)^{1/2}$$

This is denoted Sx on the 5100. In the "biased-estimate" form the standard deviation is written

$$s \cong \left[\frac{\sum \left(X - \overline{X} \right)^2}{N} \right]^{1/2}$$

and is commonly used when $N > 30$. On the 5100 it is denoted as σx.

The standard deviation is a most important property of groups because it is a quantitative measure of the risk when characterizing a group by its mean or average. For many groups or populations, 68% of the members have measurements lying within one standard deviation ($1s$) of the mean of the measurements of all the group members; 95% lie within $2s$ of the mean and 99.73% lie within $3s$ of the mean.

If two distributions have total frequencies N_1, N_2, their variations s_1^2 and s_2^2 can be combined according to the relation

$$s^2 = \frac{N_1 s_1^2 + N_2 s_2^2}{N_1 + N_2}$$

when both distributions have the same mean (not an infrequent case). The generalization of this formula to n distributions is straightforward when it is noted that the variance of n distributions with the same mean is simply the weighted arithmetic mean of the individual variances, where the weighting factor is the frequency of each distribution.

6-4 MEASURES OF DISTRIBUTION SHAPE

Before going into measures of skewness and kurtosis, we must define the moments of a distribution. The rth moment of a distribution consisting of n values is given by

$$\overline{X^r} = \frac{\sum_{j=1}^{N} X_j^r}{N} = \frac{\sum X^r}{N}$$

The first moment where $r = 1$ is the arithmetic mean. For a nonzero mean we can further define the rth moment about the mean as

$$m_r = \frac{\sum_{j=1}^{N} (X_j - \overline{X})^r}{N} = \frac{\sum (X - \overline{X})^r}{N} = \overline{(X - \overline{X})^r}$$

Similarly, the rth moment about any origin A is defined by

$$m_r' = \frac{\sum_{j=1}^{N} (X_j - A)^r}{N} = \frac{\sum (X - A)^r}{N} = \frac{\sum d^r}{N} = \overline{(X - A)^r}$$

The moments for group data can be defined in a similar manner as

$$\overline{X^r} = \frac{\sum_{j=1}^{N} f_j X_j^r}{N} = \frac{\sum f X^r}{N}$$

$$m_r = \frac{\sum f(X - \overline{X})^r}{N}$$

$$m_r' = \frac{\sum f(X - A)^r}{N}$$

The degree to which a distribution is asymmetric is specified by the skewness of the distribution. That is, if the distribution has a longer tail to the right of the distribution centroid, the distribution is said to be skewed to the right and to have a positive skewness. Conversely, if the distribution has a longer tail to the left of its centroid, the distribution is said to be skewed to the left and to have a negative skewness. Relative to the mode, the mean tends to lie on the same side as the longer tail for skewed distributions. A measure of symmetry (due to Pearson) is the difference between the mean and the mode. This difference is then divided by the standard deviation (to make the measure dimensionless) to form a measure of the skewness. Pearson's measure of skewness is

$$\text{skewness} = \frac{\text{mean} - \text{mode}}{\text{standard deviation}} = \frac{\overline{X} - \text{mode}}{s}$$

Kurtosis is a measure of the degree to which the distribution is peaked. A common measure of kurtosis is

$$a_4 = \frac{m_4}{s^4} = \frac{m_4}{m_2^2}$$

That is, the fourth moment is divided by the fourth power of the standard deviation. For Gaussian distributions, the measure of kurtosis is 3. Kurtosis is then sometimes defined by the relationship

$$a_4 - 3 = \text{kurtosis}$$

The moment of kurtosis is also referred to as the *coefficient of excess*.

6-5 PROBABILITY

Now that we have developed the basic definitions in statistics, we can examine the elements of probability, which we later relate to statistics through concepts in sampling. For our work here we define probability as follows. If an event can happen in h ways out of n equally likely ways, the probability of occurrence of this event is given by

$$p = \Pr\{E\} = \frac{h}{n}$$

we also say that this is the probability of success of the event; similarly $q = 1 - p$ is equal to the probability of failure of the event. Clearly, $p + q$ must equal 1.

If E_1 and E_2 are dependent events, that is, the probability that E_2 occurs given that E_1 has occurred is nonzero, we say that the probability E_2 will occur given E_1 is the conditional probability of E_2 given E_1 and is written

$$\Pr\{E_2|E_1\} = \frac{\Pr\{E_2 \text{ and } E_1\}}{\Pr(E_1)}$$

Now, the *multiplication law* or the law of compound probabilities (i.e., the probability that both E_1 and E_2 occur simultaneously) is given by

$$\Pr\{E_1 E_2\} = \Pr\{E_1\}\Pr\{E_2|E_1\}$$

and the *addition law* of probability (i.e., the probability that either E_1 or E_2 occurs) is given by

$$\Pr\{E_1 + E_2\} = \Pr\{E_1\} + \Pr\{E_2\} - \Pr\{E_1 E_2\}$$

If the events are mutually exclusive, then

$$\Pr\{E_1 + E_2\} = \Pr\{E_1\} + \Pr\{E_2\}$$

since $\Pr(E_1 E_2) = 0$. If the events are statistically independent, then

$$\Pr\{E_1 E_2\} = \Pr\{E_1\}\Pr\{E_2\}$$

since $\Pr(E_2|E_1) = \Pr(E_2)$.

Clearly, for events that occur discretely, such as rolling dice and flipping coins, we can form a discrete probability distribution by tabulating the event and the probability that it will occur. An example appears in Table 6-2.

Discrete probability distributions can also be plotted like histograms. Cumulative probability distributions can be developed in the same way as our cumulative relative frequency distributions. Then, as the number of probable events approaches infinity, the probability distribution functions become increasingly dense until we can form continuous probability distribution as the limit to which the discrete probability distribution approaches as the quantitization in the process approaches zero.

A concept commonly used in probability is that of expectation. Expectation is defined as follows. If X denotes a discrete random variable that can take on values $X_1, X_2, \ldots, X_k$ with associated probabilities $p_1, p_2, \ldots, p_k$,

Table 6-2   Sum of Points on a Single Throw of Two Dice

Sum of Points $= X$	Pr(X)	Cum Pr(X)
2	1/36	1/36
3	2/36	3/36
4	3/36	6/36
5	4/36	10/36
6	5/36	15/36
7	6/36	21/36
8	5/36	26/36
9	4/36	30/36
10	3/36	33/36
11	2/36	35/36
12	1/36	1

where $p_1 + p_2 + \cdots + p_k = 1$, the expectation of X is defined as

$$E(X) = \sum_{j=1}^{k} p_j X_j = \sum pX$$

The extension of this equation to continuous distributions is obvious. In this case, we would define $E(x) = \mu$(mu), the mean of the population; m (the mean of the sample) is an estimate of the true $E(x)$.

It should be clear from this discussion a very large random sample of size N from a population would result in a sample mean that is very near the population mean, whereas a sample mean based on a small sample would not necessarily be likely to be very near the population mean.

6-6 PROBABILITY DISTRIBUTIONS

Of the many distributions used in probability and statistical analysis five are frequently encountered in engineering and scientific work: the uniform, the triangular, the binomial, the Gaussian, and the Poisson distributions.

The uniform distribution is defined as

$$p(x) = \left\{ \begin{array}{ll} \dfrac{1}{b-a} & \text{for} \quad a \leqslant x \leqslant b \\ 0 & \text{for all other } x \end{array} \right\}$$

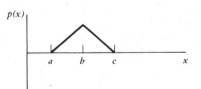

The mean value is given by

$$\mu = \frac{b-a}{2}$$

and the variance is

$$\sigma^2 = \frac{(b-a)^2}{12}$$

The triangular distribution is, as its name implies, of triangular shape and is characterized by the minimum value of its range a, the mode of the distribution b (the value of the random variable where the distribution peaks), and the maximum value of the distribution range c.

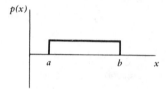

The mean value of the triangular distribution is

$$\mu = \frac{a+b+c}{3}$$

and the standard deviation is

$$\sigma^2 = \frac{(a-b)^2+(b-c)^2+(c-a)^2}{18}$$

The binomial distribution is defined by

$$p(X) = {}_NC_X p^X q^{N-X} = \frac{N!}{X!(N-X)!}p^X q^{N-X}$$

where p is the probability of success, that is, the probability that an event will happen in any single trial; q is the probability that it will fail to happen in any single trial, usually called the probability of failure and equal to $1-p$; and $p(X)$ is the probability that the event will happen exactly X times in N trials. Here X is defined only on the integers, that is, $X = 0, 1, 2, \ldots, N$. By definition this is a discrete probability distribution. Its name reflects the fact that as X takes on integer values from 1 through N, the corresponding probabilities are given by the terms in the binomial expansion

$$(p+q)^N = q^N + {}_NC_1 pq^{N-1} + {}_NC_2 p^2 q^{N-2} + \cdots + p^N$$

The statistics associated with the binomial distribution are developed in many books and will not be repeated here. Four are usually used in practical analysis, the mean, defined by

$$\mu = Np$$

the variance defined by

$$\sigma^2 = Npq$$

the coefficient of skewness defined by

$$a_3 = \frac{q - p}{(Npq)^{1/2}}$$

and the measure of kurtosis (or coefficient of excess) given by

$$a_4 = 3 + \frac{1 - 6pq}{Npq}$$

As an example of the application of the binomial distribution and its statistics, let us consider 100 flips of an unbiased coin. The probability is one-half that the coin will be heads and one-half that it will be tails. The mean thus is

$$\mu = Np$$
$$\mu = 100 \times \tfrac{1}{2} = 50$$

The standard deviation is given by the square root of the variance:

$$\sigma = \sqrt{Npq}$$
$$\sigma = \sqrt{100 \times \tfrac{1}{2} \times \tfrac{1}{2}} = \sqrt{25} = 5$$

Skewness is zero, since $p = q$, and the measure of kurtosis equals

$$\alpha_4 = 3 + \frac{1 - 6pq}{Npq}$$

$$\alpha_4 = 3 + \frac{(1 - \frac{3}{2})}{25}$$

$$\alpha_4 = 3 - \tfrac{1}{50}$$

$$\alpha_4 \approx 3$$

We see that the expected number of heads in 100 flips of a coin is 50. The standard deviation for 100 trials is 5. We would not expect the distribution of heads and tails to be skewed (if the coin is unbiased), but would expect it to be approximately Gaussian (the kurtosis of a Gaussian distribution equals 3).

The Gaussian distribution is defined by the equation

$$Y = \frac{1}{\sigma \sqrt{2\pi}} e^{-\frac{(x - \mu)^2}{2\sigma^2}}$$

where

$$\mu = \text{the mean}$$

$$\sigma = \text{the standard deviation}$$

The Gaussian distribution has the following characteristics:

1. The area bounded by the distribution and the X axis is identically equal to 1.

2. The area bounded by the distribution and the X axis in the interval between $x = a$ and $x = b$ where $a < b$ is identically equal to the probability that X lies between a and b. The Gaussian distribution is shown in Figure 6-4.

The parameters of the Gaussian distribution are the mean and variance, defined by

$$\text{mean} = \mu$$

$$\text{variance} = \sigma^2$$

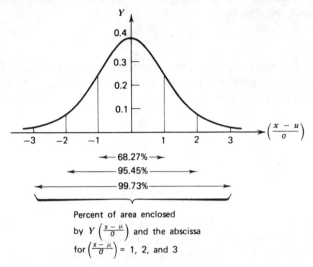

Percent of area enclosed

by $Y\left(\frac{x-\mu}{\sigma}\right)$ and the abscissa

for $\left(\frac{x-\mu}{\sigma}\right)$ = 1, 2, and 3

Figure 6-4 Gaussian distribution.

The skewness of the Gaussian distribution by inspection is zero. The measure of kurtosis for the Gaussian distribution is given by

$$a_4 = 3$$

As mentioned before, the binomial and normal distributions are related. If for a binomial random variable X, N, and P are not zero nor near zero and N is large, the binomial distribution can be approximated with the Gaussian distribution, since $(X - Np)/\sqrt{Npq}$ is approximately normal with 0 mean and unit (1) variance. It turns out that the Gaussian distribution is the limiting form of the binomial distribution as N approaches infinity. In practical work the Gaussian distribution is a reasonable approximation of the binomial distribution if both Np and Nq are greater than 5.

Another probability distribution encountered in practical probability work is the Poisson distribution, defined by

$$p(X) = \frac{\lambda^X e^{-\lambda}}{X!}$$

Here λ is a constant of the distribution.

The statistics of the Poisson distribution are all given in terms of the single parameter λ. The mean is given by

$$\mu = \lambda$$

The variance is given by

$$\sigma^2 = \lambda$$

which is identically equal to the mean (an interesting curiosity of the Poisson distribution, which we discuss later). The coefficient of skewness of the Poisson distribution is

$$a_3 = \frac{1}{\sqrt{\lambda}}$$

and the measure of kurtosis is

$$a_4 = 3 + \tfrac{1}{\lambda}$$

A number of things become apparent by examining the Poisson distribution. First, it is discrete and its basic properties are not a function of the number of trials being considered. Second, the larger the mean (λ), the less the skewness and the more the distribution tends toward the Gaussian distribution; accordingly the coefficient of excess approaches 3. From our observations of skewness and kurtosis we might expect that the Gaussian distribution is the limit to which the Poisson distribution approaches as λ approaches infinity; this, in fact, turns out to be the case.

As might be expected, the binomial and Poisson distributions are related, both being discrete and both approaching the Gaussian distribution as a limit for large-numbers samples (in the case of the binomial distribution) or a large value of the mean (in the case of the Poisson distribution). Note that in the binomial distribution if N is large and the probability p of occurrence of an event is close to zero, $q = (1-p)$ is close to 1 and we say that the event is a rare event. In these situations the binomial distribution can be closely approximated by the Poisson distribution by letting $\lambda = Np$. Comparison of the mean, variance, skewness, and coefficient of excess when $\lambda = Np$, $p = 0$, and $q \cong 1$ shows that the binomial distribution properties (e.g., mean, variance) are approximately equal to those of the Poisson distribution. Extension of the Poisson distribution to its association with the Gaussian distribution follows through its association with the binomial distribution. After a little algebra, it can be shown that the Poisson distribution can be approximated by the Gaussian distribution, since

$$\frac{(X - \lambda)}{\sqrt{\lambda}}$$

is approximately normally distributed with zero mean and unit (1) variance.

It is common in statistical analysis to use these distribution functions to model the distribution of populations being studied. The approach is to determine the mean and standard deviation of the sample of the population to estimate the mean and standard deviation of the population. The modeling process is then tested for goodness of fit, using a number of approaches. One that we discuss later in this chapter is the chi-square test. Clearly, modeling a sample of a population with the binomial or Poisson distribution solely amounts to determining the mean value $(\overline{X})$ of the sample distribution. For modeling the binomial distribution we compute

$$p = \frac{\overline{X}}{N}$$

Then

$$q = 1 - p$$

A crude test is to determine if

$$Nqp \approx s^2$$

where s is the standard deviation of the sampled population. If the calculation shows that

$$p \approx 0$$

then the Poisson distribution may better fit the data.

6-7 SAMPLING

The study of sampling deals with the determination of statistics (estimates of distribution parameters) associated with a sample drawn from a population and the population parameters themselves. We are concerned with sampling insofar as it is related to estimation, testing, and statistical inference. First we wish to define a sample, in particular a random sample. Generally it is not sufficient to sample a population in a systematic way. Since the objective is to develop the statistics of the sample and infer something about the parameters of the distribution, the sample that is representative of the population must be chosen. There are a number of methods for sampling a population, including stratified and random sampling. One universally accepted way of sampling a population so that the characteristics of the population are represented in the sample is *random sampling*. Random sampling is a process in which each member of the population has an equal chance of being included in the sample. Typically a number is assigned to each member of the population, and the numbers are then scrambled, so that each sample of the population has the same chance of being selected as any other sample.

There are two concepts of sampling: with and without replacement. The concepts are straightforward. Sampling with replacement allows each member of the population to be chosen more than once, while sampling without replacement does not. The analysis is useful because a finite sample with replacement can theoretically be considered infinite. Sampling without replacement will result in a statistic that takes into account the size of the sample compared with the total size of the population.

For any given sample of size N a value of a given statistic can be computed for that group and, if another sample of the same size is selected, in general a different value of the statistic is computed. This process can continue (with or without replacement) until all statistic values in a population are different. Clearly, if a population is of finite size and is sampled with replacement, an infinite number of values can be determined for each statistic. By organizing these values of the statistic, we obtain a distribution of the statistic itself, which is called its sample distribution. Clearly, then, there are sampling distributions of the mean, sampling distributions of the standard deviation, sampling distributions of the variance, sampling distributions of the measure of kurtosis, and sampling distributions of the coefficient of skewness. Of all these possible statistics we focus on those that help us in testing, estimating, and statistical inference only.

Suppose that we have a population of finite size, say N_p. Also suppose that a sample of size N is drawn without replacement. Then if we denote the mean of the sample distribution of the mean by

$$\mu_{\bar{X}}$$

and the standard deviation of the sample distribution of the mean as

$$\sigma_{\bar{X}}$$

and, further, if we define the population mean and standard deviation by μ and σ, respectively, we can write

$$\mu_{\bar{X}} = \mu$$

$$\sigma_{\bar{X}} = \frac{\sigma}{\sqrt{N}} \left(\frac{N_p - N}{N_p - 1} \right)^{1/2}$$

If the population is infinite or if it is finite but sampled with replacement, the population mean and standard deviations are related to the sample mean and standard deviations according to

$$\mu_{\bar{X}} = \mu$$

$$\sigma_{\bar{X}} = \frac{\sigma}{\sqrt{N}}$$

For N greater than 30, the sample distribution of the mean is approximately Gaussian with mean

$$\mu_{\text{Gaussian}} \cong \mu_{\bar{X}}$$

and standard deviation

$$\sigma_{\text{Gaussian}} \cong \sigma_{\bar{X}}$$

It is noted that the distribution of the mean of a population sample is independent of the distribution population.

Suppose now that we draw a sample from each of two populations. The number of samples drawn from the first population is N_1 and that from the second population is N_2. Now let us compute a statistic s_1. We find that each statistic has a sampling distribution, whose mean and standard deviation we denote by μ_{s_1} and σ_{s_1}. A similar situation holds true for the second population; that is, it has a mean and standard deviation given by μ_{s_2} and σ_{s_2}. If we now consider all possible combinations of these samples from the two populations, we can obtain a distribution of differences, that is, $S_1 - S_2$, which is called the sampling distribution of the differences of the statistics. The mean and standard deviation of this distribution are defined by:

$$\mu_{S_1 - S_2} = \mu_{S_1} - \mu_{S_2}$$

$$\sigma_{S_1 - S_2} = \left(\sigma_{S_1}^2 + \sigma_{S_2}^2 \right)^{1/2}$$

If S_1 and S_2 are the sample means from two populations, the sampling distribution of the differences of the means is given for infinite populations with mean and standard deviation μ_1, σ_1 and μ_2, σ_2:

$$\mu_{\bar{X}_1 - \bar{X}_2} = \mu_{\bar{X}_1} - \mu_{\bar{X}_2} = \mu_1 - \mu_2$$

$$\sigma_{\bar{X}_1 - \bar{X}_2} = \left(\sigma_{\bar{X}_1}^2 + \sigma_{\bar{X}_2}^2 \right)^{1/2} = \left(\frac{\sigma_1^2}{N_1} + \frac{\sigma_2^2}{N_2} \right)^{1/2}$$

These results hold for finite populations if sampling is with replacement. The standard deviation of a population can also be computed from a sample of the population. It, too, has a distribution with mean and standard deviations

$$\mu_s = \sigma \cong s\left(\frac{N}{N-1}\right)^{1/2}$$

$$\sigma_s = \left(\frac{\mu_4 - \mu_2^2}{4N\mu_2}\right)$$

For populations that are normally distributed this reduces to ($\mu_2 = \sigma^2$ and $\mu_4 = 3\sigma^4$):

$$\mu_s = \sigma$$

$$\sigma_s = \frac{\sigma}{\sqrt{2N}}$$

6-8 STATISTICAL ESTIMATION

We have just seen how statistical information can be computed from the data of the sampled population. One of the key problems in statistical inference is that of estimation of the population parameters (mean, variance, kurtosis, etc.) from the corresponding sample statistics. Before proceeding to the concept of confidence interval estimates for population parameters, the key issue in this section, we must clear up the issue of biased versus unbiased estimation. In computing a mean value one merely sums the numbers of the sampled values and divides by the total number. In computing the standard deviation, however, it is important to recognize that it takes at least two points to compute a variance; hence the mean of the variance of a distribution must be divided by $N-1$, not N. In this sense, then, we have an unbiased estimator. As N becomes large, the effect of biased estimation becomes less significant. For small sample sizes, however, it does matter.

Now we will estimate the *confidence interval* of population parameters. The idea here is to sample the population, then compute a mean and standard deviation for it, and try to infer what this tells us about the mean and standard deviation of the population. If we define μ_s and σ_s as the mean and standard deviation of the sampling distribution of a statistic S, then we can expect the sampling distribution of S to be approximately

Table 6-3 Two-Sided Confidence Levels Associated with σ Levels

Confidence Level (%)	σ Level z_c
50	0.6745
68.27	1.0000
80	1.28
90	1.645
95	1.96
95.45	2.00
96	2.05
98	2.33
99	2.58
99.73	3.00

Gaussian (assume that $N > 30$) and the actual sampled statistic S lying somewhere in the interval $\mu_s - \sigma_s$ to $\mu_s + \sigma_s$, or $\mu_s - 2\sigma_s$ to $\mu_s + 2\sigma_s$, or $\mu_s - 3\sigma_s$ to $\mu_s + 3\sigma_s$; about 68.3%, 95.5%, and 99.7% of the time, respectively.

In other words, we can be confident of finding μ_s in the interval $S - \sigma_s$ to $S + \sigma_s$ about 68.3% of the time; or in the interval $S - 2\sigma_s$ to $S + 2\sigma_s$ about 95.5% of the time; or in the interval between $S - 3\sigma_s$ to $S + 3\sigma_s$ about 99.7% of the time. We can say that we are 68.3% confident that the mean lies somewhere in the interval $S \pm \sigma_s$, that we have 95.5% confidence that the mean lies somewhere in the interval $S \pm 2\sigma_s$; and that we have 99.7% confidence that the mean μ_s lies somewhere in the interval $S \pm 3\sigma_s$. The relationship between confidence levels and σ levels is shown in Table 6-3.

The formula for computing the confidence interval associated with the mean is given by

$$\overline{X} \pm z_c \sigma_{\overline{x}}$$

If the sampling is from an infinite population or from a finite population but with replacement, the confidence limits in the estimate of the mean are specified by

$$\overline{X} \pm z_c \frac{\sigma}{\sqrt{N}}$$

if sampling is without replacement from a population of finite size N_p:

$$\bar{X} \pm z_c \frac{\sigma}{\sqrt{N}} \left(\frac{N_p - N}{N_p - 1} \right)^{1/2}$$

The equations for confidence intervals for differences and sums of two statistics S_1 and S_2 are given by

$$(S_1 - S_2) \pm z_c \sigma_{S_1 - S_2} = (S_1 - S_2) \pm z_c \left(\sigma_{S_1}^2 + \sigma_{S_2}^2 \right)^{1/2}$$

$$(S_1 + S_2) \pm z_c \sigma_{S_1 + S_2} = (S_1 + S_2) \pm z_c \left(\sigma_{S_1}^2 + \sigma_{S_2}^2 \right)^{1/2}$$

provided that the distribution of S_1 and S_2 is approximately Gaussian.

By way of example, note that the confidence limits for the difference of two populations means in a case where the populations are infinite or are finite but with replacement are given by

$$\left(\bar{X}_1 - \bar{X}_2 \right) \pm z_c \sigma_{\bar{X}_1 - \bar{X}_2} = \left(\bar{X}_1 - \bar{X}_2 \right) \pm z_c \left(\frac{\sigma_1^2}{N_1} + \frac{\sigma_2^2}{N_2} \right)^{1/2}$$

The confidence interval for standard deviations of a normally distributed population is given by

$$S \pm z_c \sigma_S = S \pm \frac{z_c \sigma}{\sqrt{2N}}$$

Occasionally we need to reference probable error; we define it here but retain the concept for reference. The 50% confidence limits of the population parameters corresponding to a statistic S are given by $S \pm 0.6745\sigma_s$. This quantity is known as the probable error of the estimate and may be worth memorizing.

6-9 SAMPLING IN THE SMALL

Earlier we made use of the fact that there are simplifications in the formulas for computing a statistic when $N > 30$. Here we are concerned with cases when N is substantially less than 30, and the emphasis is on the determination of the statistics associated with small samples and on the distribution of those statistics. Specifically, we consider here the Student's t-distribution and the chi-square distribution.

The Student's t-distribution is as follows:

$$t = \frac{\overline{X} - \mu}{s} \sqrt{N - 1} = \frac{\overline{X} - \mu}{s/\sqrt{N - 1}}$$

If we consider samples of size N selected from a Gaussian distribution with mean μ and if we compute t given the sample mean and sample standard deviation s, the sampling distribution for t can be obtained. This distribution is given by

$$Y = \frac{Y_0}{\left(1 + t^2/(N-1)\right)^{N/2}} = \frac{Y_0}{\left(1 + \dfrac{t^2}{\nu}\right)^{(\nu+1)/2}}$$

Here Y_0 is a constant depending on N and is such that the area under the t-distribution is 1. The constant $\nu = (N-1)$ is called the number of degrees of freedom. This distribution is called "Student's" t-distribution. Note that for large values of ν or N ($N > 30$) the curves closely approximate the Gaussian distribution:

$$Y = \frac{1}{\sqrt{2\pi}} e^{-t^2/2}$$

We can define the 95 and 99% confidence intervals by either computing the confidence intervals on the pocket calculator or using a table of t-distributions. Specifically, if $-t_{0.975n} + t_{0.975}$ are the values of t for which 2.5% of the area lies in each tail of the t-distribution, then a 95% confidence level for t is

$$-t_{0.975} < \frac{\overline{X} - \mu}{s} \sqrt{N - 1} < t_{0.975}$$

Then, μ is expected to lie in the interval

$$\overline{X} - t_{0.975}\left(\frac{s}{\sqrt{N-1}}\right) < \mu < \overline{X} + t_{0.975}\left(\frac{s}{\sqrt{N-1}}\right)$$

with 95% confidence.

In general, we can represent the confidence limits for population means by

$$\overline{X} \pm t_c \frac{s}{\sqrt{N-1}}$$

6-10 CHI-SQUARE

We now proceed to the chi-square random variable, which is defined by

$$\chi^2 = \frac{Ns^2}{\sigma^2} = \frac{\left(X_1 - \overline{X}\right)^2 + \left(X_2 - \overline{X}\right)^2 + \cdots + \left(X_N - \overline{X}\right)^2}{\sigma^2}$$

The chi-square distribution is defined by

$$Y = Y_0(\chi^2)^{1/2(\nu-2)} e^{-(\chi^2/2)} = Y_0 \chi^{\nu-2} e^{-\chi^2/2}$$

Here $\nu = N - 1$ is the number of degrees of freedom and Y_0 is a constant depending on ν such that the total area under the curve is 1. The chi-square distribution corresponding to various values of ν are shown in Figure 6-5. As was done with the normal and t-distributions, we can

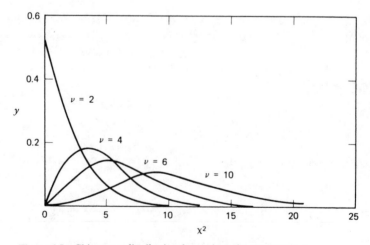

Figure 6-5 Chi-square distribution for various degrees of freedom.

define 95 and 99% or other confidence limits and intervals for chi-square by using a table of the chi-square distribution. In this manner we can estimate, within specified confidence limits, the population standard deviation σ in terms of the sample standard deviation s. If $\chi^2_{0.025}$ and $\chi^2_{0.095}$ have the values of χ^2 for which 2.5% of the area lies in each tail of the distribution, the 95% confidence interval is

$$\chi^2_{0.025} < \frac{Ns^2}{\sigma^2} < \chi^2_{0.975}$$

We can see that σ is estimated to be in the interval

$$\frac{s\sqrt{N}}{\sqrt{\chi^2_{0.975}}} < \sigma < \frac{s\sqrt{N}}{\sqrt{\chi^2_{0.025}}}$$

with 95% confidence.

6-11 STATISTICAL FORECASTING TECHNIQUES

Linear regression is a name given to fitting a straight line through a set of data relating two random variables. The straight line establishes a *trend* between the two variables, hence is frequently called a trend line. When one of the variables is time, a trend line refers to the trend of another variable over a period of time. The intervals between the time periods are usually fixed, but this need not be the case. In fact, to fit a curve to a set of data points requires no restriction on the intervals between the variables.

The objective is to fit a line (the equation for a line is $y = A + Bx$) through a set of data pairs (y and x). The problem is to determine A and B from the set of ys and xs input into the calculator. When the user inputs

$$
\begin{array}{ccc}
y_1 & \text{and} & x_1 \\
y_2 & \text{and} & x_2 \\
y_3 & \text{and} & x_3 \\
\vdots & & \vdots \\
y_n & & x_n
\end{array}
$$

the 5100 will automatically compute A and B so that the line

$$y = A + Bx$$

best* passes through the point pairs (y_i, x_i). The calculator calculates A

*Best in the sense of minimizing the sum of the square of the variance of each y_i from $y = A + Bx$ at x_i over the range of i from zero to n.

with the equation

$$A = \frac{\Sigma y \Sigma x^2 - \Sigma x \Sigma xy}{n\Sigma x^2 - (\Sigma x)^2}$$

and B with the equation

$$B = \frac{n\Sigma xy - \Sigma x \Sigma y}{n\Sigma x^2 - (\Sigma x)^2}$$

Straight-line forecasting is useful for short-range considerations. What about long-range forecasts of exponential growth or decline? In these cases it would be better to try to fit the exponential curve

$$y = be^{mx}$$

to the data, not $y = A + Bx$. Taking the logarithm of the exponential equation yields

$$\ln(y) = \ln(b) + (mx)\ln(e)$$

Since $\ln(e) = 1$, we see that

$$\ln(y) = \ln(b) + mx$$

If we let $\ln(b) = A$ and $B = m$, we can rewrite this equation in the form

$$\ln(y) = A + Bx$$

which is the same form as the straight-line equation. Thus if we input the variables $\ln(y)$ and x instead of y and x, we can calculate a straight-line curve fit through $\ln(y)$. Then in the last step, calculate the antilog of $\ln(y)$ to get

$$y = be^{mx}$$

Example 6-1   If 20% of the transistors produced by a process are defective, using the binomial distribution determine the probability that out of four transistors chosen at random, one, none, and at most two will be defective. Since the probability of a defective transistor is

$$P = .2$$

and of a nondefective transistor it is

$$q = 1 - p = .8$$

then

probability (1 defective transistor out of 4)

$$= {}_4C_1(.2)^1(.8)^3 = \frac{4!}{1!(4-1)!}(.2)^1(.8)^3 = .4096$$

This problem is solved on the 5100 using the AER mode, by inputting the algebraic equation

$$P = {}_NC_x P^x q^{N-x}$$

as

Here

$$A = N$$
$$B = x$$
$$C = P$$

Then we quickly find

probability (0 defective transistors) $= {}_4C_0(.2)^0(.8)^4 = .4096$

probability (2 defective transistors) $= {}_4C_2(.2)^2(.8)^2 = .1536$

Thus

probability (at most 2 defective transistors)

= probability (0 defective transistors)

+ probability (1 defective transistor)

+ probability (2 defective transistors)

$= .4096 + .4096 + 1536 = .9728$

Example 6-2 Numerically integrate the Gaussian probability density function form $z = -2.1$ to 0.30 to determine the probability that the random variable z will occur on the interval $[-2.1$ to $0.30]$ and where

$$z = \frac{x - \mu}{\sigma}$$

$$Y = \frac{1}{\sigma\sqrt{2\pi}} e^{-z^2/2}$$

$$\mu = 0$$

$$\sigma = 1$$

Approach:

Using Simpson's extended rule, we will numerically integrate Y with the formula

$$I = \int_0^{4\Delta z} Y(z)\, dx \simeq \frac{\Delta z}{3}(Y_0 + 4Y_1 + 2Y_2 + 4Y_3 + Y_4)$$

Analysis:

Equating the range of the interval with the upper bound of the integral we find that

$$4\Delta x = (0.30 - (-2.1)) = 2.4$$

Thus

$$\Delta x = 0.6$$

Now, Y is evaluated with the 5100 at the ends of the interval with the program:

$$1; \quad f(A) = e - (A^2 \div 2) \div \sqrt{(2 \times \pi)}$$

Then we can calculate the five values of Y required for the numerical integration as:

$$Y_0 = Y(-2.1) = 0.0440$$

$$Y_1 = Y(-1.5) = 0.1295$$

$$Y_2 = Y(-0.9) + 0.2661$$

$$Y_3 = Y(-0.3) = 0.3814$$

$$Y_4 = Y(0.3) = 0.3814$$

Substituting these values into the numerical integration formula we find

$$I \simeq 0.6(0.0440 + 4 \times 0.1295 + 2 \times 0.2661 + 4 \times 0.3814 + 0.3814) \div 3$$
$$I \simeq 0.6002$$

Thus the probability that a random sample of z will fall on the interval $[-2.1 \text{ to } 0.3]$ is $\simeq 60.02\%$. Another way to say this is that $\simeq 60.02\%$ of a large sample of x will be found on the interval

$$2.1 \leqslant \frac{x-\mu}{\sigma} \leqslant 0.3$$

Finally, the exact value of I is 0.60000. We see then that our answer is accurate to one tenth of 1%. This example was selected to illustrate a rule of thumb the author has found useful for engineering work:

On the interval $[0; 4\sigma]$; the four-interval extended Simpson rule

$$P \simeq \frac{\Delta z}{3}(Y_0 + 4Y_1 + 2Y_2 + 4Y_3 + Y_4)$$

where

$$\Delta z = \frac{subinterval \text{ of } [0; 4\sigma]}{4}$$

is accurate enough to calculate single-sided Gaussian probabilities to one tenth of 1%.

Example 6-3 The mean weight of 500 engineers in 170 lb, and the standard deviation is 15 lb. Assuming the weights to be normally distributed, find how many engineers weigh between 139 and 174 lb. Weights recorded as being between 139 and 174 lb can actually have any value from 138.5 to 174.5 lb, assuming that they are recorded to the nearest pound.

$$138.5 \text{ lb in standard units} = \frac{138.5 - 170}{15} - -2.10$$

$$174.5 \text{ lb in standard units} = \frac{174.5 - 170}{15} - 0.30$$

Then the number of engineers between 139 and 174 lb is the area under the Gaussian probability distribution curve between $z = -2.1$ and $z = 0.03$ multiplied by 500 (see Figure 6-6). The area can be calculated by numerically integrating the function

$$Y = \frac{1}{\sigma\sqrt{2\pi}} e^{-\frac{(x-\mu)^2}{2\sigma^2}}$$

from $z = (x - \mu)^2/2\sigma^2 = -2.1$ to $z = 0.3$ or by table look-up. It is found that the area equals 0.6.

The number of engineers weighing between 139 and 174 lb is

$$500(0.6000) - 300.$$

Example 6-4 Find the probability of obtaining between 3 and 6 heads inclusive in 10 tosses of a coin, first by using the binomial distribution and then by using the Gaussian distribution approximation of the binomial distribution. The binomial distribution of Figure 6-7 gives

$$\text{probability (3 heads)} = {}_{10}C_3\left(\tfrac{1}{2}\right)^3\left(\tfrac{1}{2}\right)^7 = \tfrac{15}{128}$$

$$\text{probability (4 heads)} = {}_{10}C_4\left(\tfrac{1}{2}\right)^4\left(\tfrac{1}{2}\right)^6 = \tfrac{105}{512}$$

$$\text{probability (5 heads)} = {}_{10}C_5\left(\tfrac{1}{2}\right)^5\left(\tfrac{1}{2}\right)^5 = \tfrac{63}{256}$$

$$\text{probability (6 heads)} = {}_{10}C_6\left(\tfrac{1}{2}\right)^6\left(\tfrac{1}{2}\right)^4 = \tfrac{105}{512}$$

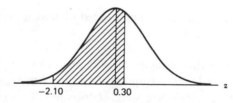

Figure 6-6 Percent of engineers weighing between 139 and 174 lb.

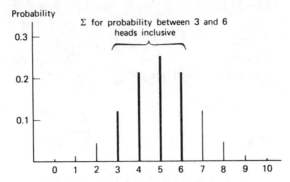

Figure 6-7 Discrete binomial distribution.

Thus

probability (of getting between 3 and 6 heads inclusive)
$$= 15/128 + 105/512 + 63/256 + 105/512 = 99/128 = 0.7734$$

This problem is solved on the 5100 using the AER mode in an iterative way, building on the formula developed in Example 1. In this case we let B increase by one on each press of the *COMP* key and let *P* accumulate the results in memory location D as follows:

1; D = D + A¢B × CYˣB × (1 − C)Yˣ(A − B) STO D, B = B + 1 STO B

With

A initially 10 (use 10 STO A to initialize)
B initially 3 (use 3 STO B to initialize)
C initially 0.5
D initially 0

The probability distribution for the number of heads in 10 tosses of the coin is shown graphically in Figure 6-8. The required probability is the sum of the areas of the cross-hatched rectangles and can be approximated by the area under the corresponding normal curve shown dashed.

Using the normal distribution, 3 to 6 heads can be considered as 2.5 to 6.5 heads. The mean and variance for the binomial distribution are given by

$$\mu = Np = 10\left(\tfrac{1}{2}\right) = 5$$

and

$$\sigma = \sqrt{Npq} = \sqrt{(10)\left(\tfrac{1}{2}\right)\left(\tfrac{1}{2}\right)} = 1.58$$

$$2.5 \text{ in standard units} = \frac{2.5 - 5}{1.58} = -1.58$$

and

$$6.5 \text{ in standard units} = \frac{6.5 - 5}{1.58} = 0.95$$

The probability of getting between 3 and 6 heads inclusive in the area between $z = -1.58$ and $z = 0.95$ under the normal distribution curve is

(area between $z = -1.58$ and $z = 0$) + (area between $z = 0$ and $z = 0.95$)

$$= 0.4429 + 0.3289 = 0.7718$$

which compares very well with the true value 0.7734 obtained using the binomial distribution.

Example 6-5   Measurements of the diameters of a random sample of 81 Pitts Special engine-mount bolts showed a mean diameter of $\tfrac{5}{8}$ in. and a standard deviation of 0.0005 in. Find (*a*) 95% and (*b*) 99% confidence limits for the mean diameter of all the bolts in the lot.

The 95% confidence limits are given by

$$\overline{X} \pm \frac{1.96\sigma}{\sqrt{N}}$$

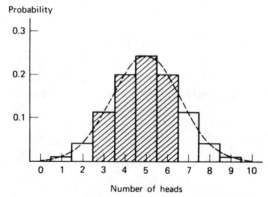

Figure 6-8   Data treated as continuous normal distribution.

which is approximately

$$\overline{X} \pm \frac{1.96s}{\sqrt{N}}$$

In this example, $\overline{X} = 0.625$ in. and $s = 0.0005$ in. Thus

$$0.625 \pm 1.96\left(\frac{0.0005}{\sqrt{81}}\right)$$

$$0.625 \pm 0.000109$$

The 95% confidence interval is then (0.624891, 0.625109).

The 99% confidence limits are given by

$$\overline{X} \pm \frac{2.58s}{\sqrt{N}}$$

or 6.25 ± 0.000143. Thus one can expect that 99 times out of 100 the diameter of a bolt selected at random is on the interval (0.62486, 0.62514). Note that we have assumed the standard deviation to be the unbiased standard deviation.

Example 6-6 A small sample of 10 measurements of the same Pitts Special engine-mount bolts gives a mean of $\overline{X} = 0.625$ and a standard deviation of $s = 0.0004$. Again find the 95% and 99% confidence limits for the bolt diameter, taking into account the very small sample size.

In this case the 95% two-sided confidence limits are the same as the 97.5% single-sided confidence limits and are given by

$$X \pm t_{0.975}\left(\frac{s}{\sqrt{N-1}}\right)$$

Here

$$\nu = 9$$

so we find

$$t_{0.975} = 2.26$$

from Table 6-4 or directly using the definition of the t distribution. Using $\overline{X} = 0.625$ and $s = 0.0004$, we can be 95% confident that the actual mean will be included in the interval

$$0.625 \pm 2.26\left(\frac{0.0004}{\sqrt{10-1}}\right) = 0.625 \pm 0.000301 \text{ in.}$$

**Table 6-4 Percentile Values (t_p) for the Single-Sided
t-Distribution with v Degrees of Freedom**

v	$t_{0.995}$	$t_{0.99}$	$t_{0.975}$	$t_{0.95}$	$t_{0.90}$
1	63.66	31.82	12.71	6.31	3.08
2	9.92	6.96	4.30	2.92	1.89
3	5.84	4.54	3.18	2.35	1.64
4	4.60	3.75	2.78	2.13	1.53
5	4.03	3.36	2.57	2.02	1.48
6	3.71	3.14	2.45	1.94	1.44
7	3.50	3.00	2.36	1.90	1.42
8	3.36	2.90	2.31	1.86	1.40
9	3.25	2.82	2.26	1.83	1.38
10	3.17	2.76	2.23	1.81	1.37
11	3.11	2.72	2.20	1.80	1.36
12	3.06	2.68	2.18	1.78	1.36
13	3.01	2.65	2.16	1.77	1.35
14	2.98	2.62	2.14	1.76	1.34

The 99% confidence limits are given by

$$\overline{X} \pm t_{0.995}\left(\frac{s}{\sqrt{N-1}} \right)$$

where $v = 9$ and $t_{0.995} = 3.25$. Then with 99% confidence we can expect the actual mean to be included in the interval

$$0.625 \pm 3.25\left(\frac{0.0004}{\sqrt{10-1}} \right) = 0.625 \pm 0.000433 \text{ in.}$$

We previously worked this problem assuming that large sampling methods are valid and now we can compare the results of the two methods. In each case the confidence intervals obtained by using the small or exact sampling methods are greater than those obtained by using large sampling methods even though the standard deviation of the smaller sample was less than the standard deviation of the larger sample.

PROBLEM SOLVING REFRESHER FOR THE 5100 *

A-1 FOUR ARITHMETIC CALCULATIONS

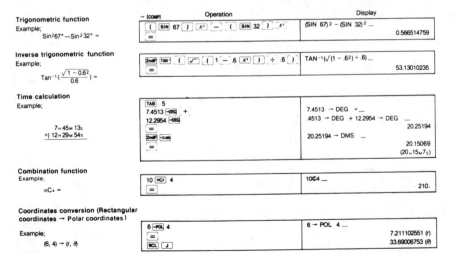

Mixed calculation
Example;
$$2300 \times (-24) \div 0.12 =$$

→ [COMP] Operation	Display
2300 [×] [(-)] 24 [÷] .12	2300 × −24 ÷ .12 _
[=]	−460000.

$$\frac{(54 \times 10^5 + 6.76 \times 10^6)}{1.25 \times 10^{-12}} =$$

[(] 54 [Exp] 5 [+] 6.76 [Exp] 6 [)]	(54E5 + 6.76E6) _
[÷] 1.25 [Exp] [(-)] 12	(54E5 + 6.76E6) ÷ 1.25E −12 _
[=]	9.728E 18

Memory calculation
Example;
$$46 + 78 + 61 =$$
$$-) \ 423 - 154 + 26 =$$
$$+) \ 72 + 86 + 45 =$$

[CL] [=M]	0.
46 [+] 78 [+] 61 [M+]	185.
423 [−] 154 [+] 26 [2ndF] [M+]	295.
72 [+] 86 [+] 45 [M+]	203.
[RM]	93.

A-2 FUNCTION CALCULATIONS

Trigonometric function
Example;
$$\text{Sin}^2 67° - \text{Sin}^2 32° =$$

→ [COMP] Operation	Display
[(] [SIN] 67 [)] [x²] [−] [(] [SIN] 32 [)] [x²]	(SIN 67)² − (SIN 32)² _
[=]	0.566514759

Inverse trigonometric function
Example;
$$\text{Tan}^{-1}(\frac{\sqrt{1 - 0.6^2}}{0.6}) =$$

[2ndF] [TAN⁻¹] [(] [√] [(] 1 [−] .6 [x²] [)] [)] [÷] .6 [)]	TAN⁻¹(√(1 − .6²) ÷ .6) _
[=]	53.13010235

Time calculation
Example;

$$7_H 45_M 13_S$$
$$+) \ 12_H 29_M 54_S$$

[TAB] 5	
7.4513 [→DEG] [+]	7.4513 → DEG + _
12.2954 [→DEG]	.4513 → DEG + 12.2954 → DEG _
[=]	20.25194
[2ndF] [→DMS]	20.25194 → DMS _
[=]	20.15069
	(20ₕ15ₘ7ₛ)

Combination function
Example;
$$_{10}C_4 =$$

10 [nCr] 4	10C4 _
[=]	210.

Coordinates conversion (Rectangular coordinates → Polar coordinates)

Example;
$$(6, 4) → (r, θ)$$

6 [→POL] 4	6 → POL 4 _
[=]	7.211102551 (r)
[RCL] [J]	33.69006753 (θ)

* Better copy was not available at press time.

A-3 LINEAR REGRESSION

	Height	Weight
A	172cm	67kg
B	167	54
C	179	68
D	163	51
E	181	70
F	173	59
G	169	61

r = coefficient of correlation

$\left.\begin{array}{l}a = \\ b = \end{array}\right\}$ coefficient of regression
(y = a + bx)

A person 162cm would weigh?

Operation	Display
→ [STAT]	STAT MODE
[2ndF] [CA]	0.
172 [x.y] 67 [Data]	1.
167 [x.y] 54 [Data]	2.
179 [x.y] 68 [Data]	3.
163 [x.y] 51 [Data]	4.
181 [x.y] 70 [Data]	5.
173 [x.y] 59 [Data]	6.
169 [x.y] 61 [Data]	7.
[2ndF] r	0.904970862
[2ndF] a	− 115.4657375
[2ndF] b	1.028455285
[2ndF] a +	− 115.4657375 + _
[2ndF] b × 162	4657375 + 1.028455285 × 162 _
=	51.14401867

A-4 COSINE THEOREM

Formula;

$$a = \sqrt{b^2 + c^2 - 2bc \cos\theta}$$

b = B
c = C
θ = D

When b = 2, c = 3 and
θ = 60° → a = ?

a = 1; ANS 1

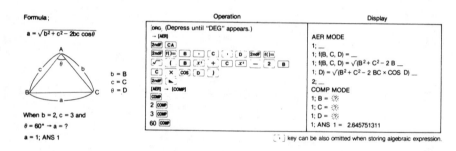

Operation	Display
[DRG] (Depress until "DEG" appears.)	
→ [AER]	AER MODE
[2ndF] [CA]	1; _
[2ndF] f()= B , C , D [2ndF] f()=	1; f(B, C, D) = _
√‾ (B x² + C x² − 2 B	1; f(B, C, D) = √(B² + C² − 2 B _
C × cos D)	1; D) = √(B² + C² − 2 BC × COS D) _
[2ndF] ◄	2; _
[AER] → [COMP]	COMP MODE
[COMP]	1; B = ░
2 [COMP]	1; C = ░
3 [COMP]	1; D = ░
60 [COMP]	1; ANS 1 = 2.645751311

[·] key can be also omitted when storing algebraic expression.

A-5 THE CENTER OF GRAVITY OF A TRIANGLE

Formula;

$$y_G = \frac{y_1 + y_2 + y_3}{3} \qquad x_G = \frac{x_1 + x_2 + x_3}{3}$$

x₁ = A,
x₂ = B,
x₃ = C

y₁ = D,
y₂ = E,
y₃ = F

When x₁ = 1, x₂ = 2 and x₃ = 4 → x_G = ?
y₁ = 1, y₂ = 3, and y₃ = 2 → y_G = ?
x_G = 1; ANS 1   y_G = 2; ANS 1

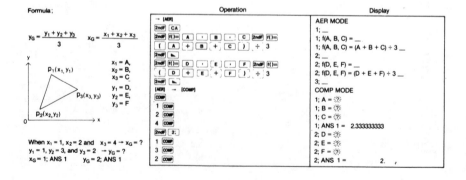

Operation	Display
→ [AER]	AER MODE
[2ndF] [CA]	1; _
[2ndF] f()= A , B , C [2ndF] f()=	1; f(A, B, C) = _
(A + B + C) ÷ 3	1; f(A, B, C) = (A + B + C) ÷ 3 _
[2ndF] ◄	2; _
[2ndF] f()= D , E , F [2ndF] f()=	2; f(D, E, F) = _
(D + E + F) ÷ 3	2; f(D, E, F) = (D + E + F) ÷ 3 _
[2ndF] ◄	3; _
[AER] → [COMP]	COMP MODE
[COMP]	1; A = ░
1 [COMP]	1; B = ░
2 [COMP]	1; C = ░
4 [COMP]	1; ANS 1 = 2.333333333
[2ndF] 2;	2; D = ░
1 [COMP]	2; E = ░
3 [COMP]	2; F = ░
2 [COMP]	2; ANS 1 = 2.

A-6 IMPEDANCE OF A SERIES CIRCUIT (RLC SERIES CIRCUIT)

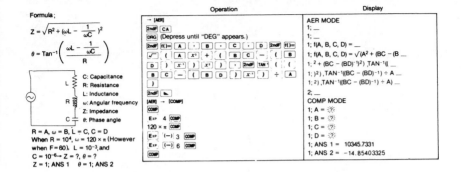

A-7 MACH NUMBER FOR A JET PLANE

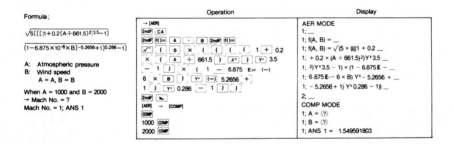

A-8 MOVEMENT OVER A FLAT SURFACE INVOLVING AN INCREASING SPEED

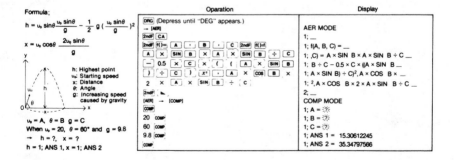

INDEX